Rural Land Degradation in Australia

Other Meridian titles:

Jamie Kirkpatrick, *A Continent Transformed: Human Impact on the Natural Vegetation of Australia*

Robert H Fagan and Michael Webber, *Global Restructuring*

David Chapman, *Natural Hazards*

Clive Forster, *Australian Cities: Continuity and Change*

Rural Land Degradation in Australia

Arthur and Jeanette Conacher

Series editors
Deirdre Dragovich
Alaric Maude

Melbourne
OXFORD UNIVERSITY PRESS
Oxford Auckland New York

OXFORD UNIVERSITY PRESS AUSTRALIA

Oxford New York
Athens Auckland Bangkok Bombay
Calcutta Cape Town Dar es Salaam Delhi
Florence Hong Kong Istanbul Karachi
Kuala Lumpur Madras Madrid Melbourne
Mexico City Nairobi Paris Singapore
Taipei Tokyo Toronto
and associated companies in
Berlin Ibadan

OXFORD is a trade mark of Oxford University Press

First published 1995

National Library of Australia
Cataloguing-In-Publication data:

Conacher, A.J. (Arthur J.) 1939–.
Rural land degradation in Australia.
Bibliography.
Includes index.
ISBN 0 19 553436 0
1. Environmental degradation—Australia. 2. Land use,
Rural—Australia. 3. Soil degradation–Australia. 4. Soil
conservation—Australia. 1. Conacher, Jeanette. II.
Institute of Australian Geographers. III. Title.
(Series: Meridian, Australian geographical perspectives).
333.761370994

Indexed by Arthur and Jeanette Conacher
Cover photograph: Effects of land use on land degradation
near Narrogin in the Western Australian wheatbelt.
Source: Western Australian Department of Agriculture.
Typeset by Dobell, Temple & Associates
Printed by SRM Production Services Sdn. Bhd., Malaysia
Published by Oxford University Press
253 Normanby Road, South Melbourne, Australia

Foreword

Australian geographers have produced some excellent books in recent years, several of them in association with the 1988 bicentennial of European settlement in the continent, and all of them building on the maturing of geographical research in this country. However, there is a continuing need for relatively short, low-cost books written for university students to fill the gap between chapter-length surveys and full-length books, to explore the geographical issues and problems of Australia and its region, or to present an Australian perspective on global geographical processes.

Meridian: Australian Geographical Perspectives is a series initiated by the Institute of Australian Geographers to fill this need. The term meridian refers to a line of longitude linking points in a half-circle between the poles. In this series it symbolises the interconnections between places in the global environment and global economy, which is one of the Key themes of contemporary geography. The books in the series cover a variety of physical, environmental, economic and social geography topics, and are written for use in first and second year courses where the existing texts and reference books lack a significant Australian perspective. To cope with the very varied content of geography courses taught in Australian universities the books are designed not as comprehensive texts, but as modules on specific themes which can be used in a variety of courses. They are intended to be used either in a one-semester course, or in a one-semester component of a full year course.

Titles in the series will cover a range of topics representing contemporary Australian geographical teaching and research, such as economic restructuring, vegetation change, natural hazards, the changing nature of cities, land degradation, gender and geography, and urban environmental problems. Although the emphasis in the series is on Australia, we also intend to produce some titles on Southeast Asia, using the considerable expertise that some Australian geographers have developed on this region.

We hope geography students will find the series informative, lively and relevant to their interests. Individual titles will also be of interest to students in related disciplines, such as environmental sciences, planning, economics, women's studies and Asian Studies.

Arthur and Jeanette Conacher's book is the fourth in the series. It examines land degradation—a group of processes which are a major threat to the sustainability of Australian environments, farming systems and rural communities. The authors provide an Australia-wide examination of the causes, extent and consequences of land degradation. They go beyond the usual issues of soil erosion, soil fertility decline and dryland salinity to emphasise the significance of vegetation degradation and the loss of genetic resources, as well as the degradation of surface and groundwater. They argue that land degradation is an ecological problem, and show the relationships between different forms of land degradation and also the interactions between causes and effects. Consequently there is a chapter on the effects of agricultural chemicals on land and people, because these chemicals are both a response to land degradation and a cause of further problems.

Arthur and Jeanette Conacher have been actively involved in environmental research and environmental issues for a number of years. Their book is based on a wealth of experience, and readers will find it informative, interesting and practical. It provides a very readable and accessible examination of the scientific understanding of land degradation processes, of the social, economic and political causes and consequences of this group of environmental problems, and of some of the solutions to them. We are confident that the book, like others in the *Meridian* series, will be of interest and value to both students and the general reader.

Deidre Dragovich
University of Sydney
Alaric Maude
Flinders University

Contents

Foreword v
List of plates xi
List of figures xii
List of tables xiv
Acknowledgments xvi

1 *The rural land degradation problem* 1

What is land degradation? 2
What is in this book? 3
Global land degradation 6
Land degradation is not new 10
Nature and extent of land degradation in Australia 12
Forms of land degradation 14
Increased extent of degradation since 1975 14
Land degradation forms not covered by the 1975 survey 16

2 *Degradation of ecosystems* 19

Clearing of forests and woodlands 19
Degradation of woodlands and grasslands 22

Ecological disturbance 24
Extinct, rare and endangered species 25
Loss of genetic resources 31
Introduced species 32
Weeds 33
Other pests 36
Conclusion 37

3 *Problems associated with the use of agricultural chemicals* 38

Adverse effects of synthetic fertiliser applications 38
Water repellency 39
Soil acidity 39
Induced deficiencies 40
Heavy metals 40
Soil organisms 41
Eutrophication 41
Nitrates 44
Adverse effects of using synthetic pesticides 45
Toxicity of pesticides 46
Pesticide and veterinary chemical use in Australia 47
Pesticides in the environment 48
Minimum tillage and pesticides 48
Pest resistance to pesticides 52
Spray drift 52
Residues in food 53
Pesticide residues and human health 54
Conclusion 55

4 *Economic and social implications of land degradation* 58

Costs of repairing land degradation 58
Production losses 62
Costs of pesticides 65
Indirect costs 65

Agricultural expansion 65
Some consequences of irrigation farming 67
Increasing farming intensity 69
Off-farm land degradation 69
Water quality 69
Land-use conflicts 75
Competition for agricultural land 75
Indirect land-use conflicts 76
Rural population decline 79
Declining country towns 80
Social implications 81

5 *Direct causes of rural land degradation* 83

Clearing and ecological properties 84
Weeds 85
Fire and ecosystem stress 86
Nutrient cycling 86
Water movements 89
Effects of replacing native vegetation with introduced plants 93
Accelerated erosion 95
Subsurface water 97
Secondary salinisation and waterlogging 100
Cultivation and soil properties 100
Wind erosion 101
Conclusion 103

6 *Underlying causes of land degradation* 104

Economic pressures 105
The nature of the Australian environment 112
People's attitudes 113
Pioneer attitudes and the rural ethos 116
Farmer education and resistance to change 117
Lack of awareness 118
Government agencies 119
Government policies 121
Conclusion 122

7 *Solutions to land degradation problems* 124

Loss of habitat and protecting ecosystems 125
Protecting genetic diversity 128
Diminishing the adverse effects of agricultural chemicals 128
Pesticides 129
Fertilisers 130
Veterinary chemicals 130
Avoidance of synthetic chemicals 130
Dealing with soil degradation 131
Other approaches to better management 134
Agroforestry 134
Whole-farm and integrated catchment planning 136
Soil Conservation Districts and Landcare 137
Ecologically sustainable land management 139
Who pays? 141
Conclusion 142

References 144
Index 160

Plates

1.1	Degraded land, Spain	10
1.2	Erosion on the loess plateau, China	11
5.1	'Litter dams' under eucalypt forest	98–9
6.1	Treeless wheatbelt paddock	114

Figures

1.1	Global soil degradation	7
1.2	General land use, boundary of the arid zone, and proportion of land in use requiring some form of treatment in 1975–77, by state	13
1.3	Estimated extent and cost of land degradation problems in Western Australia	18
3.1	Occurrence of blue-green algal blooms in New South Wales during 1991/92	43
3.2	Agricultural and veterinary chemicals, index of quantities sold in real terms in Australia 1975–88	47
3.3	Prices paid for selected farm inputs in Australia 1967/68–1979/80	49
3.4	Growth in the area of crops treated with direct drill herbicide in Western Australia 1971–84	50
3.5	Farm chemicals responsible for ill health	55
3.6	Symptoms of reported ill-effects from farm chemical use	57
4.1	Wheat yields per hectare since 1960 for Australia and selected OECD countries	63
4.2	Irrigated regions in the Murray valley	73
4.3	River Murray salinity	74
5.1	Diagrammatic illustration of an undisturbed pedon with its natural vegetation	87

5.2 Loss of soil nutrients in surface soil materials following forest clear felling 90
5.3 Diagrammatic soil pedon showing lateral water movements and aquifers 94
5.4 Diagrammatic illustration of the effects of vegetation on wind erosivity 101
5.5 Relationship between soil movement by wind and biomass of the vegetation cover 102
5.6 Relationship between wind erosion and tillage 103
6.1 Increase in areas under crop and pasture 1950–90 106
6.2 Prices received/prices paid by farmers 1960–90 107
6.3 Total institutional farm debt 1976–91 108
6.4 Australia: number of agricultural establishments 1952–92 109
6.5 Australian farm employment, including employers, self-employed, wage and salary earners and unpaid family helpers 110
6.6 Numbers of new tractors sold in Australia, by type, 1976–92 111
7.1 Geometric principles for the design of nature reserves 126
7.2 The growth of Landcare groups in Australia, by state and territory, from 1985 139

Tables

1.1 Annual economic losses caused by desertification 8
1.2 Annual costs of preventing, correcting and rehabilitating desertification 9
1.3 Analysis of forms of degradation in areas requiring treatment at June 1975 15
1.4 Land degradation in New South Wales, 1988 17
2.1 Loss of forests and woodlands, expressed as a percentage of the areas covered by forests and woodlands at the time of European settlement 20
2.2 Areas ('000 hectares) of native forest and woodland in the Australian states and territories in 1990 21
2.3 Total number of major plant alliances per structural formation in each state of Australia, and the Northern Territory, and the percentages of these alliances not recorded or poorly represented in existing reserves 23
2.4 Numbers of plant species by endangered status and state 27
2.5 Extinct, endangered and vulnerable vertebrate species in Australia 28–30
2.6 Some major weed species in Australia 34–5
3.1 Consumption of fertilisers in terms of elemental NPK 39
3.2 Comparison of available pesticides, restrictions and cancellations in Victoria and the US, 1990–91 56–7

4.1 Summary of treatment measures required and construction costs of works at June 1975 59

4.2 Estimated expenditure by state/territory soil conservation agencies in 1989/90 60

4.3 Commonwealth contribution to soil conservation and Landcare, 1974/75 to 1990/91 61

4.4 Average per farm business expenditure on farm chemicals by industry, 1991–92; also indicating these as proportions of total farm costs, and with some Canadian comparisons 66

4.5 Problems likely to affect water resources to the year 2000 71

4.6 Examples of tolerance to salt concentrations in water for selected plants, animals and human activities 72

4.7 Yield losses of some agricultural plant species exposed to O_3 under various experimental conditions 79

5.1 Amounts of selected nutrients in trees, shrubs, litter and soils on a red earth site near Pemberton, southwestern Australia 88

5.2 Some examples of median values of canopy interception as a percentage of annual or seasonal gross precipitation 91

5.3 Comparison of volumes of water per area generated by overland flow and stemflow on forested slopes in southwestern Australia 92

6.1 List of issues considered by the populations surveyed by Newman and Cameron, in order of importance 115

7.1 Areas set aside for conservation purposes, by states and territories, in 1988 125

7.2 Benefits of agroforestry in semi-arid conditions 135

Acknowledgments

The following are acknowledged for granting permission to use copyright material: Commonwealth of Australia, Australian Government Publishing Service, for tables 1.3, 2.1, 2.3, 2.4, 2.5, 4.1, 4.2, 4.3 and 4.5, and figure 4.3; Commonwealth Department of Primary Industries and Energy for figure 7.2; CSIRO Office of Space Science and Applications for table 7.1; The Kondinin Group, Belmont, Western Australia, for figures 3.5 and 3.6; Morescope Publishing for table 3.1 and figure 6.2; OECD, Paris, for table 4.7; the Office of Environment, Victoria, for figures 5.5 and 5.6 and table 3.2; Western Australian Department of Agriculture, Media Services Section, for the cover photograph. The authors and the publisher would be pleased to hear from those copyright holders who have not yet replied to our written inquiries. The diagrams and maps were drawn by Di Milton and Guy Foster, whose work is acknowledged with gratitude. The photographs were printed by the University of Western Australia's Media Services; and the librarians at Reid and Biological Sciences libraries, University of Western Australia and the Library and Information Service of Western Australia, were particularly helpful. Hilary Bailey assisted cheerfully with typing the tables. Except for the cover photograph, all photographs were taken by Arthur Conacher.

1

The rural land degradation problem

We abuse land because we regard it as a commodity belonging to us. When we see land as a community to which we belong, we may begin to use it with love and respect.

(Leopold, 1949: Foreword)

The outlook for succeeding generations is indeed dismal should the destruction of the forests continue as in the past; our watersheds will become bare, bald hills, from which torrential floods will devastate the alluvial plains. . . . The preservation of our indigenous flora, whilst looked upon as a fad by the ignorant and unthinking, is really in its cumulative effects one of great national importance.

(Samuel Dixon, paper to the Royal Society of South Australia, 1892; cited in Hawke, 1989:1)

For more than 40000 years, Australia supported at least 1200 generations of humankind, providing life in that time to an estimated 360 million Australians as hunter-gatherers. Although this was not without its impact on the landscape, particularly through the Aboriginal use of fire, European agriculturalists in just five generations have altered the face of the Australian continent far more dramatically than had been done by all those who had gone before. Our imprints on the land are clearly visible from space.

(modified from Graetz et al., 1992:v)

WHAT IS LAND DEGRADATION?

Definitions are uncomfortable things: so often they seem only to make something apparently quite straightforward much more complicated. 'Land degradation' is no exception, unfortunately. But an attempt to define it is essential, to ensure understanding of what is being considered in this book and, just as importantly, what is not. One of the fundamental difficulties is distinguishing between the *nature* and *process* of land degradation, its *causes* and the *effects*. We are not entirely happy that we have succeeded in making these distinctions throughout this book as they are inextricably interwoven. Additionally, land degradation is a complex response to the interaction between people and their biophysical environment.

The first question that needs to be answered is: what is meant by 'land'? When farmers buy land, they do not just buy the soil and perhaps the topography. The 'land' also includes the indigenous and exotic plants and animals on the property, the soil biota, the insects and the birds, the springs, streams and groundwater (their presence adds value), and the aesthetics—which are not covered in this book. Arguably it also includes air—certainly it does in the micro-climatic sense, and farmers, American Indians and ourselves would include it—but its incorporation here as an element of 'land' would add too many complications. Clearly, however, air pollution, the hole in the ozone layer, the enhanced Greenhouse effect and other climatic phenomena have important implications for land degradation and require some attention. Legally, in Australia, a farmer's land does not include the underlying minerals—though farmers would probably dispute that too—and we have excluded consideration of the effects of mining.

Using 'land' in the above, broad sense reflects the United Nations Environment Programme's (UNEP, 1993:5) definition of it as a concept which '*includes soil and local water resources, land surface and vegetation or crops that may be affected by one or a combination of processes acting on it*'. To this we would explicitly add fauna, as indicated above.

It follows that 'degradation' of land refers to a decline in value of all the above elements of the biophysical environment; but note carefully that 'value' is not measured solely or even primarily in economic or utilitarian terms—that is, in terms of its usefulness to people. Degradation can also indicate a loss of something intrinsic to that object. For example, the functioning of an ecosystem—through some interruption to energy flows or food chains, perhaps, or disruption of the habitat of some animal or bird—may be 'degraded' even though it has no directly adverse consequence for people.

There is one widely accepted, additional element in defining 'land degradation'—namely that it results from human actions.

These concepts of land and its degradation reflect those being used in a number of the European Union's MEDALUS projects (Mediterranean Desertification and Land Use) and by the International Geographical Union's Study Group on Desertification in Regions of Mediterranean Climate (MED). The latter group recently adopted the following working definition of land degradation, as proposed by A.J. Conacher (February 1994): '*alterations to all aspects of the natural (or biophysical) environment by human actions, to the detriment of vegetation, soils, landforms and water (surface and subsurface, terrestrial and marine) and ecosystems*.'

Defining land degradation as resulting from human actions explicitly draws attention to the fact that the range of environmental problems termed 'natural hazards' is excluded. Natural hazards include phenomena such as earthquakes, tsunamis, volcanic eruptions, tropical cyclones and, less clearly, floods, droughts and bushfires. The difficulty with the latter is that they are often exacerbated, if not actually caused, by human actions; and where this is the case they should be included under the term 'land degradation'. Thus, insofar as 'desertification' (defined by UNEP as land degradation in arid and semi-arid regions) is caused by human actions, such as overclearing of vegetation or overstocking of domestic animals, and not just natural processes, then it is clearly a land degradation problem. The astute reader will note a circular argument developing here. Nevertheless, it should be noted that natural hazards may themselves cause or intensify land degradation.

WHAT IS IN THIS BOOK?

While recognising that land degradation is a global problem, and that it is considerably more severe in a number of countries other than Australia, the focus here is on Australia, with the next section briefly placing Australia in a global context. It may be comforting to observe that Australia's problems are generally relatively minor in comparison with those of a number of other countries, but this is no cause for complacency. The recency of European settlement and farming practices in this country means that our problems have had a relatively short time in which to develop (although Aborigines used fire, they did not engage in agricultural or pastoral activities). On the other hand, explosive population growth associated with widespread adoption of technological farming, allied with consumerism, are equally recent phenomena throughout the world.

Degradation of agricultural areas—extensive cropping and livestock farming, and horticulture—is emphasised, while rangelands (the 'pastoral' country) are given less attention. This is not to say that rangelands do not have problems of degradation, and additional information concerning rangeland degradation and management may be found in Messer and Mosley (1983), Butcher (1992), Grant (1992) and National Rangeland Management Working Group (1994), among others.

Problems in the coastal zone, which is not usually considered 'rural' (even if the coast in question is a long way from the nearest city), are excluded from this book. Degradation within urban areas—even where these affect 'natural' features such as indigenous bushland, or river reserves—are also not treated here, although this should not be seen as diminishing their importance. On the other hand, off-site impacts of urban areas on, for example, air and water quality in rural areas receive some attention.

Land degradation is sometimes seen in relatively narrow terms—referring to soil erosion by water and wind, and perhaps salinisation and the decline of some soil physical properties. Global and Australia-wide surveys of land degradation and desertification have tended to reflect this, although even the out-of-date 1975 survey of land degradation in Australia (discussed later in this chapter) included vegetation degradation, as did the more recent Western Australian assessment. The remainder of chapter 1 presents this rather traditional approach to the topic.

But weeds, for instance, are a multi-million dollar problem in Australia, yet they are rarely recognised as such. People discuss the loss of species—species extinctions are highly topical—but not invasions of exotics, although the latter is commonly discussed for fauna. To deal with weeds and other pests, farmers reach for their chemicals; and more farmers make more passes over their broadacre fields dispensing pesticides than they do when ploughing. The former is a far more important practice, with its consequences, than the widely recognised ploughing. Thus, some attention is given later in the book to the use of synthetic chemicals in agriculture.

Our approach is quite deliberately not the usual, prosaic, 'checklist' on land degradation: many publications are available which regurgitate the same data, using the same format and the same, tired old arguments. We have tried not only to inform but also to challenge the reader into thinking more broadly about the dynamics and complexities of land degradation and the problems of data, and especially incorporating the idea of land degradation as an ecological problem. This is now reflected, in Australia, by the 'whole farm' and 'integrated catchment management' approaches to managing the land.

Thus the remainder of this chapter looks first at global and then Australian land degradation, briefly, in a fairly traditional way, ending with a discussion of what has *not* been assessed. Chapter 2 considers some of those aspects not covered by the more traditional approach, looking at land degradation more as an ecological problem. The implications of farmers' responses to an out-of-kilter ecosystem, of using agricultural chemicals, are considered in chapter 3. A prime reason for concern about land degradation is the social and economic implications of the problem, and these are discussed in chapter 4.

By the end of that chapter it is hoped that some realisation of the complexities of the problem will have been attained, and in particular an appreciation that we are not dealing merely with 'soil erosion'. It is then appropriate to consider causes: first the more obvious, direct causes (chapter 5) and then the underlying context, which includes the cost/price squeeze, the nature of the Australian environment, peoples' attitudes, perceptions and awareness of the problems, and the roles of government agencies and policies (chapter 6). Without that context, the problem of land degradation makes little sense. Finally, it is important to recognise that many people have reached a good understanding of various aspects of land degradation, and that a considerable amount of work is being done to solve the problems. Chapter 7 canvasses some of the solutions that are being implemented by a range of people and organisations.

Before continuing, it is appropriate to comment briefly on the reason for geographers' interest in the issue of land degradation. The intimate connection between human activities and the natural environment, plus the fact that these interactions take place in a spatial context—and vary markedly from place to place—means that land degradation is intrinsically geographical in nature and of prime interest to geographers. However, the problem also has a temporal component; many problems (such as the salinisation of irrigated lands in the ancient civilisations of North Africa, Mesopotamia and the Indus valley) have occurred previously, whereas others (such as pollution of agricultural lands by industrial wastes) are much more recent. Virtually all the problems are connected to the unrelenting growth in numbers of human beings, and their growing demands on the physical resource base, making the search for solutions through a clear understanding of causes increasingly urgent.

This, too, makes study of land degradation particularly appropriate for geographers. In recent years, geographers have become dissatisfied with 'merely' searching for explanations, difficult and challenging though that may be: today, many geographers want to use their knowledge and expertise to help make the world a better place in which to live. However, the word 'help' must be stressed. Many skills are involved in understanding and developing solutions to land degradation. In addition to geography, the disciplines most involved include soil science, botany, hydrology, zoology and ecology on the one hand, and political science, law, economics and sociology on the other. Those geographers who integrate the two sides of their discipline, the human and the physical, have a distinctive and particularly valuable perspective; but they should not make the mistake of believing they can do it on their own.

GLOBAL LAND DEGRADATION

Land degradation is a global problem. The global store of agricultural land continues to decline through urbanisation, unsustainable agricultural practices and deforestation; and a significant portion of the remaining arable and grazing land is under considerable pressure from compaction by livestock and farm implements, over-use of fertilisers and pesticides, salinisation, alkalisation and acidification, depletion of nutrients, water and wind erosion, and deterioration of drainage (Tolba et al., 1992:132).

According to a recent fifteen-year global assessment of soil degradation, 15 per cent of the world's land area has been degraded by human activities. Of the areas affected, more than half (55%) were due to water erosion, nearly a third (28%) to wind erosion, about 12 per cent to chemical soil degradation (including salinity) and 4 per cent to physical interference causing waterlogging, compaction and subsidence (Anon., 1992). The worst-affected areas are the world's drylands, which cover more than 40 per cent of the world's terrestrial surface. Here, land degradation as a result of human activity (unsustainable agricultural practices and deforestation) and exacerbated by natural events such as climatic variations, is called 'desertification' (UNEP, 1992).

Figure 1.1 shows the extent of global soil degradation. Extensive areas of the world are affected—especially the margins of the Sahara in Africa, parts of the so-called 'Middle East' (which is actually to the north and west of the Australian continent), a band across southern and eastern Europe, large areas in northern China and Mongolia, parts of Australia (although not particularly severely affected), and areas in the south-west of North America, Central and South America, and southern Africa.

The more arid regions contain almost one billion people or a sixth of the world's population, with some of the world's highest birth rates (Tolba et al., 1992). The people living in these areas are both the agents and the victims of land degradation. As many as half of them have experienced at some stage the direct or indirect effects of desertification.

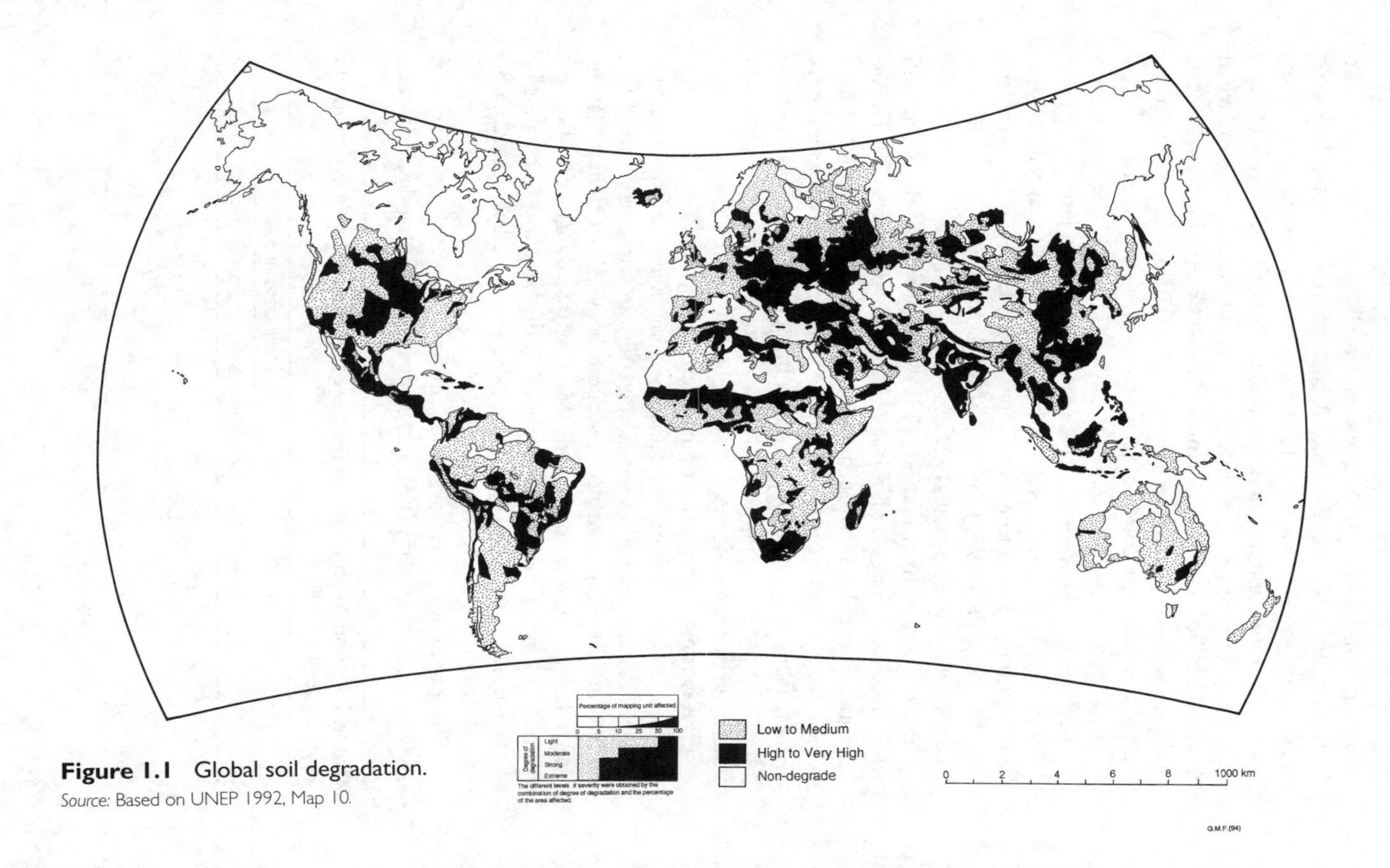

Figure 1.1 Global soil degradation.

Source: Based on UNEP 1992, Map 10.

Manifestations include chronic hunger, eye diseases, bloated stomachs, high mortality rates, massive rural depopulation, debilitated livestock, empty grain storage bins, sand-blasting of crops, sediment-filled reservoirs, dust storms and destructive floods. In 1990, direct economic losses (average annual income forgone) due to desertification were estimated at US$42 billion (table 1.1). Annual costs of preventative, corrective and rehabilitative methods range between US$10 and $22 billion (table 1.2), and indirect costs are between two and ten times greater (Dregne et al., 1991). These data may be compared, on the one hand, with World Bank development assistance funding to developing countries in 1989 of US$34 billion (UNDP, 1991), and on the other with annual global military expenditure which approached US$1000 billion in the late 1980s (SIPRI, 1990), or annual OECD agricultural subsidies of US$300 billion (Tolba et al., 1992).

In 1977, the United Nations (UN) estimated that the world will have lost close to one-third of its arable land by the year 2000. The organisation pointed out that such a loss during a period of unprecedented population growth and increased demands for food could be disastrous, as it is, at the time of writing in 1993, in much of Africa south of the Sahara. Yet, at least two recent findings run counter to the picture of unrelieved gloom and doom.

One is research reported by Hellden (1991). Hellden and others analysed satellite and airborne remote sensing imagery of the Sahel region in Africa, supported by ground investigations, from the early 1970s to 1987 and found no evidence to support the widely held view that the desert is encroaching southwards. Most major changes could be explained by varying rainfall characteristics, with boundaries fluctuating by 200 km from dry to wet years (UNEP, 1992; see also Olsson (1993), who writes about the 'myth' of desertification).

The other was a report from the United Nations Food and Agriculture Organisation (FAO) in September 1992 that despite famine

Table 1.1 Annual economic losses caused by desertification in the world's drylands (billion $US).

	Annual income forgone	*Annual cost of anti-desertification measures*			
		Preventative	*Corrective*	*Rehabilitation*	*Total*
Irrigated croplands	10.8	0.5 – 1.6	0.9 – 2.5	1.0 – 2.0	2.4 – 6.1
Rain-fed croplands	8.2	0.6 – 1.8	0.9 – 2.8	1.1 – 3.0	2.7 – 7.5
Rangelands	23.3	0.3 – 0.9	0.7 – 1.9	2.0 – 6.0	5.0 – 8.8
Total drylands	42.3	1.4 – 4.2	2.4 – 7.2	6.2 – 11.0	10.0 – 22.4

Source: Tolba et al., 1992: table 3.

Table 1.2 Annual costs of preventing, correcting and rehabilitating desertification in the world's drylands (billion $US).

	Preventive	*Corrective*	*Rehabilitation*	*Total*
Total global cost	1.4 – 4.2	2.4 – 7.2	6.2 – 11.0	10.0 – 22.4
Cost to 18 countries not requiring assistance	0.6 – 1.8	1.0 – 3.0	2.4 – 3.0	4.0 – 7.8
Cost to 81 countries requiring external assistance	0.8 – 2.4	1.4 – 4.2	3.8 – 8.0	6.0 – 14.6

Source: Tolba et al., 1992: table 2.

in Somalia and the threat of widespread food shortages across Africa, the number of hungry people in developing countries *declined* by 155 million over the previous two decades (*The West Australian*, 22 Sept. 1992). The FAO attributed the improvement to better seeds, cropping techniques and government food policies, and noted that average food availability rose from 2290 calories per person per day in 1961–63 to 2700 calories in 1988–90—a nutritionally adequate amount (FAO, 1991).

But these figures need to be treated with caution. Regional discrepancies still occur: a nutritional gap of almost 1000 calories per person exists between and within the developing countries. The increase in per capita food availability in developing countries slowed in the 1980s compared with the 1960s and 1970s, and for some areas such as sub-Saharan regions, actually worsened (World Food Council, 1991). In fact, it is not widely realised that food output per person, per year, peaked some years ago in all major regions of the world: in 1967 in Africa, 1978 in Eastern Europe and the (previous) Soviet Union, 1981 in Latin America and North America, and 1984 in Western Europe, Asia and the world as a whole (Brown, 1991: table 1-2).

The FAO also pointed out that 786 million people are underfed, including 200 million children who suffer malnutrition. A high proportion of these are the rural poor. In Europe and Africa, malnutrition is such a problem among an estimated 35 million refugees that diseases like scurvy and beriberi have reappeared; and about 2 billion people worldwide still lack key nutrients in their diet, which can lead to serious diseases and retardation. These problems are not solely attributable to desertification or land degradation: wars, in particular, play a significant role. But it is clear that land degradation is a major cause. A recent World Bank Working Paper has noted that the alleviation of poverty in developing countries is closely linked to environmental sustainability (Goodland and Daly, 1993).

Land degradation is not new

Historical evidence shows that land degradation (and desertification) is not a new phenomenon. Extensive land degradation occurred centuries ago in four regions: around the Mediterranean sea, in the Mesopotamian valley, on the loess plateau of China, and in central America.

Land degradation around the Mediterranean was caused as long ago as 2600 BC by deforestation, with timber being used for temple and ship-building in Egypt, for example, and elsewhere by centuries of clearing for agriculture and overgrazing, mostly by goats (plate 1.1). Even in more temperate regions of Western Europe, population pressures combined with climatic variability caused some severe regional problems in agriculture by the late Middle Ages (Butzer, 1974; Blaikie and Brookfield, 1987). The lower valleys of the Tigris and Euphrates in the region known as Mesopotamia have the most significant—and almost certainly most ancient—instances of land degradation caused by waterlogging, sedimentation and salinisation of irrigated lands (Jacobsen and Adams, 1958). Salinity problems became serious for the first time around 2400 BC in the Girsu area of southern Iraq. That phase lasted until about 1700 BC and was followed by a less serious period of salinisation in central Iraq about 3000 years ago. The last major period of salinisation began 800 years ago east of Baghdad. Evidently, some of the long-abandoned fields recovered after water tables were lowered, but other areas, especially in the south, have never recovered their previous productivity.

Plate 1.1 Degraded land following centuries of grazing by goats, Spain.

Another major degraded area is the loess plateau of semi-arid China, in and around the big bend of the Hwang He (Yellow River—so named for its massive loads of yellow, eroded silt particles from the loess). Near Lanzhou, a large industrial city on the Hwang He, the loess reaches its maximum thickness of 318 m: it is derived from wind-blown, silt-sized (0.02–0.002 mm diameter) particles transported from the deserts to the north, particularly the Gobi. The loess is highly susceptible to erosion by water (plate 1.2) and has been subjected to accelerated erosion—ranging from massive gullying by surface runoff, to extensive subsurface tunnel-gully erosion, to huge landslides—during the thousands of years of human occupation, materially assisting the construction of China's vast coastal plain and deltas. Dregne (1986) refers to one estimate that gullies comprise 26 000 km^2 of the 600 000 km^2 of loess lands.

The huge erosion rates on the loess plateau, in association with changing river flows, are having more recent, serious implications downriver as well as on the plateau itself. River flows in the Hwang He have decreased significantly in recent decades, as increasing populations and demands for agricultural produce have led to increased use of both surface and groundwaters for irrigation (Smil, 1992; Douglas, 1992). This diversion of water from the Hwang He, which amounts to more than a quarter of the total flow in dry years, has reduced silt transport to Bohai

Plate 1.2 Massive, deep-seated gullying and mass movement on the loess plateau, Jen Chou valley experimental rehabilitation area, China. At this site the local population has been relocated, terracing has been constructed and a variety of trees and shrubs planted to stabilise the materials. Reductions of up to 90% in erosion rates have been claimed as a result of these practices.

Bay. Up to a quarter of the river's annual sediment load of about 400 Mt is now deposited on the river bed in Henan and Shandong provinces. As a result the river bed has been rising about one metre per decade. Here, the river is confined between about 1400 km of dykes, which are between 3 and 15 m above the surrounding countryside, and which protect about 250 000 km^2 of the North China Plain. A breach of the dykes south of Jinan, in the most vulnerable area, would flood up to 33 000 km^2, affecting 18 million people (roughly Australia's total population) (Smil, 1992).

In a fourth region of the world, Central America, population pressures on the land are among the explanations for the collapse of the classical Mayan culture. Clearing of tropical forests, cropping and urban construction led to accelerated soil erosion, evidenced by substantial lacustrine sedimentation (Deevy et al., 1979; Bunney, 1990). Recovery may be slow. In northern Mexico, soils have remained degraded almost 900 years after agriculture ceased in the region (Holliday, 1988).

NATURE AND EXTENT OF LAND DEGRADATION IN AUSTRALIA

The most comprehensive source of data on the nature and extent of land degradation in Australia is still the survey conducted in 1975–77 by the Department of Environment, Housing and Community Development (1978). These data were made somewhat more accessible by Woods (1983), who presented much of the information in a series of maps with a greater emphasis on regions. The two 1986 volumes by the Department of Arts, Heritage and Environment (1986a, b) on the *State of the Environment in Australia* relied on the 1975–77 data for information on land degradation, although this was supplemented by new data on inland waters, forests and woodlands, and flora and fauna. In 1989, a federal inquiry into land degradation in Australia noted that the 1975–77 study was still being widely quoted (even the Prime Minister's important *Statement on the Environment* in 1989 used these data—Hawke, 1989) and that attempts to obtain better estimates had been hampered by the lack of an agreed methodology. Although it was recognised that the 1975–77 data were out of date, several witnesses to the inquiry considered that the situation had not improved and that the 1975–77 data probably represented a *minimum* estimate of the magnitude of remedial costs (House of Representatives . . . , 1989).

Indeed, according to the Department of Arts, Heritage and Environment's (1986) reports on the State of the Environment, only 2 per cent of agricultural and pastoral land identified in 1975–77 as being in need of soil conservation measures had been treated. Further, since that time, the country has experienced several severe droughts, cropping has expanded into more fragile, marginal areas, and there have been substantial increases in the extent of land affected by salinity, acidity and soil structure decline. Although some of these problems have been offset by better management practices, it is likely that the net effect has been an increase in the extent of degraded land.

The overall findings of the 1975–77 survey are presented in figure 1.2. Of particular interest is the huge discrepancy between the figure for

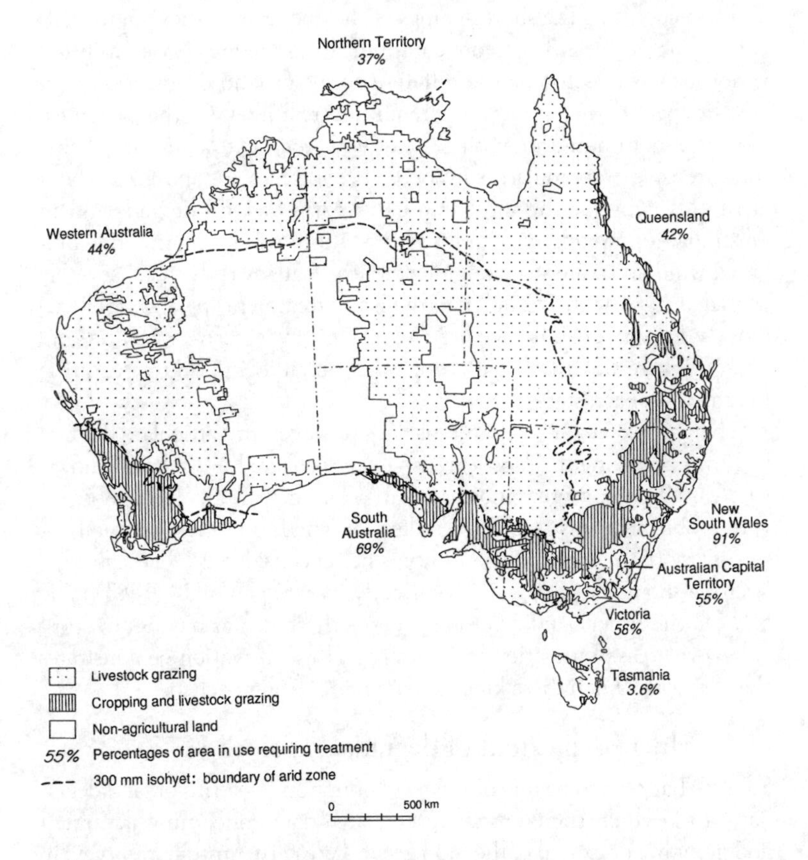

Figure 1.2 General land use of Australia, boundary of the arid zone, and proportion of land in use requiring some form of treatment in 1975–77, by state.

Source: Compiled from data in Woods (1983), with land use data from the Division of National Mapping (1982).

New South Wales, with 91 per cent of the land in use requiring treatment of some kind in order to repair land degradation, and the figure for Tasmania (3.6%). While one can only speculate as to the reasons for this discrepancy, it does call into question the accuracy of the data.

Of particular importance is the fact that the data are *estimates*, often based on subjective assessments, obtained from soil conservation and agricultural officers in the various states and territories. The data were not obtained from accurate surveys.

Forms of land degradation

While recognising the shortcomings of the above data, they are worth a closer look. Table 1.3 presents data indicating the areas of land in the various states affected by the different forms of land degradation covered by the 1975–77 survey, and requiring treatment. Of the total area, most occurs in non-arid grazing areas, with the next largest area being the extensive cropping areas (Woods, 1983: tables 4.1 and 4.2). Water erosion was by far the most extensive form of land degradation in Australia, followed by vegetation degradation, wind erosion, and combined wind and water erosion. New South Wales was the most severely degraded state, with 262 000 km^2 requiring treatment in 1975, followed closely by Queensland with 225 000 km^2. Western Australia and Victoria were roughly equal third, with about 100 000 km^2 requiring treatment in each state.

There are some odd data in table 1.3, however. In particular, Victoria is shown as having the most extensive areas affected by dryland salinity, whereas all other sources indicate that Western Australia has a more severe problem in this regard (regardless of whether primary (natural) or secondary (human-induced) salinity is being considered). The figure for Victoria should be 1500 km^2, as correctly shown in Woods' (1983) table 8.1. On the other hand, Victoria is correctly shown as having the most extensive areas of irrigation area salinity. This observation demonstrates the care with which these kinds of data must be approached.

Increased extent of degradation since 1975

Salinity happens to be one of the few—if not only—forms of land degradation for which there are comprehensive data concerning its extent and growth. For example, the most recent Western Australian survey, in 1989, revealed that the extent of secondary, dryland salinity in that state had nearly doubled since 1975, to 4400 km^2 (George, 1990).

Somewhat older data from Queensland also indicate that the area of land requiring treatment has increased since 1975 (Department of Arts,

Table 1.3 Analysis of forms of degradation in areas requiring treatment at June 1975.

Form of degradation	*Australia*	*NSW*	*Vic*	*Qld*	*SA*	*WA*	*Tas.*	*NT*	*ACT*
					'000 km²				
Area in use	1804	303	168	780	130	215	26	180	1.1
Area not requiring treatment	987	41	69	525	94	112	25	120	0.30
Water erosion	577	199	58	198	17	70	0.83	34	0.77
Wind erosion	57	–	26	–	18	13	0.02	–	–
Combined wind and water erosion	55	41	–	0.41	–	14	0.08	–	–
Vegetation degradation	92	8.1	–	57	–	2.7	–	24	–
Dryland salinity sometimes in combination with water erosion	9.7	–	6.5	(a)	0.56	2.7	–	–	–
Irrigation area salinity	9.0	0.6	8.3	–	0.11	–	–	–	–
Other	14	13	0.42	0.06	0.011	0.7	–	–	–
Total areas requiring treatment	815	262	99	225	36	103	0.93	58	0.77

All values are approximate only and have therefore been rounded to two or three significant figures depending on the need for precision and the accuracy of the estimates.

(a) Dryland salinity was reported in seven of the 24 land zones delineated in the Queensland non-arid areas for the study, but the area affected and the treatment measures required were not assessed.

Source: Woods, 1983: table 5.1. © Commonwealth of Australia; reproduced by permission.

Heritage and Environment, 1986b: table 1.3). Of all regions in the state, only the Darling Downs recorded a *decrease* (of 41 000 ha) in the 'area of cultivated land requiring soil conservation measures' from 1977 to 1982: increases in all the other regions totalled 379 000 ha (to a total of 2.335 million ha), ranging from 4000 ha in South Burnett to 203 000 in Capricornia. It should also be noted that, over the same period, the area *protected* by soil conservation measures increased by 242 000 ha to 933 000 ha.

Land degradation forms not covered by the 1975 survey

There are other reservations concerning the 1975 data reported by Woods (1983). In particular, not all forms of land degradation were covered by the 1975–77 survey. Other problems relating to soils, for example, include: waterlogging, soil acidification, loss of organic matter, loss of soil nutrients, increasing water repellency, loss of soil structure (including the formation of plough pans and subsoil hardpans), reduction of soil biota, and accumulation of excess nutrients or residues of heavy metals and pesticides; as well as offsite costs of degradation.

Detailed and reliable data relating to the whole range of different forms of rural land degradation have been scarce until recently, and still are. Examples of recent studies follow.

The first example is a report produced as a result of a New South Wales land degradation survey conducted in 1987/88 (New South Wales Soil Conservation Service 1989). A relatively detailed breakdown of types of land degradation is presented in table 1.4. Wind erosion and loss of soil structure are perhaps the most important problems in that state, while mass movement, surface scalding and especially invasion by woody shrubs are also significant problems. Of these, only wind erosion was covered by the 1975 survey.

The comprehensive *1991 State of the Environment Report* for Victoria found that over 70 per cent of all agricultural land is experiencing moderate to severe soil structural decline, and more than half is affected by severe soil acidity. These two problems are not recent, and they rank well ahead of wind and water erosion (Scott, 1991).

Some detailed assessments of land degradation in the Murray/Darling drainage basin (encompassing parts of New South Wales, Victoria and South Australia) have been presented in various Murray/Darling studies (Murray/Darling Basin Commission 1989; Murray/Darling Basin Ministerial Council, 1987a, b; 1989).

In Western Australia, the Department of Agriculture provided 1989 estimates to a parliamentary inquiry both of areas affected and production

Table 1.4 Land degradation in New South Wales, 1988.

	Severity	*% state affected*	*Area (km²)*
Water erosion	moderate	1.81	14510
sheet and rill	severe	0.53	4250
	v. severe	0.31	2520
gully	moderate	5.68	45500
	severe	4.80	38490
	v. severe	0.67	5400
	extreme	0.07	570
Mass movement of slopes	present	2.90	
Wind erosion	moderate	14.06	112700
	severe	9.04	72460
	v. severe	1.57	12550
Saline seepage	moderate	0.54	
	severe	0.66	
Irrigation salinity	moderate	0.52	
	severe	0.25	
Surface scalding	moderate	9.18	73570
	severe	0.88	7060
Soil acidification	moderate	6.65	53300
	severe	3.38	27120
Soil structure loss	moderate	6.48	51910
	severe	10.86	87010
Woody shrub invasion	moderate	12.75	102160
	severe	4.41	35330

Source: Cocks, 1992: table 2.4.

losses in relation to some problems of land degradation, and those data are presented in figure 1.3. Leaving aside vegetation decline in the rangelands, which occurs primarily in the arid zone and is not included in the table 1.3 data, it can be seen that problems of subsoil compaction, water repellence, soil structure decline and waterlogging all exceed water erosion (the major problem reported in the 1975–77 survey), and that the above problems in addition to soil acidification are also more important than wind erosion (see also Grant, 1992).

It is clear that the findings from the 1975–77 national survey, which was carried out to provide a basis for soil conservation policy in Australia, are outdated. It is not sufficient to advocate that they be treated with caution; the time has come to abandon those data and to conduct another Australia-wide survey in order to provide up-to-date, accurate and more comprehensive data on the problems of land degradation in Australia. Moreover, there are other forms of degradation in addition to soil deterioration and vegetation decline.

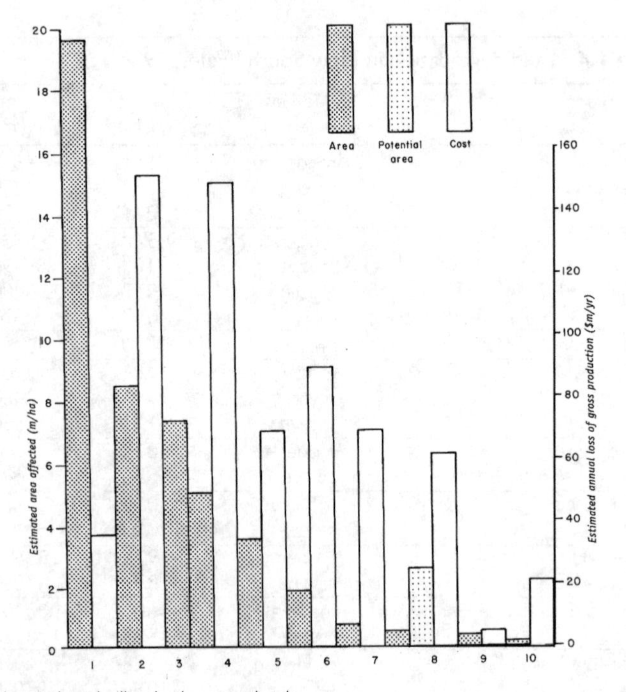

1 Vegetation decline in the rangelands
2 Subsoil compaction
3 Wind and water erosion in the rangelands (part of 1)
4 Water repellence
5 Soil structure decline
6 Waterlogging (crop: 0.5 m ha; pasture: 1.3 m ha: ave. yrs)
7 Water erosion
8 Dryland salinity
9 Soil acidification
10 Wind erosion (variable)

Figure 1.3 Estimated extent and cost of land degradation problems in Western Australia.

Source of data: Select Committee 1991, table 1.

Despite the serious shortcomings of the 1975–77 survey, it is probable that more than two-thirds of Australia's agricultural land—and 50 per cent of the pastoral areas—are affected to some degree by one or more forms of land degradation as a consequence of human actions. Some land requires urgent remedial attention, and more extensive areas need significant modifications to land management practices in order to prevent the current extent and severity of degradation from intensifying.

2

Degradation of ecosystems

Forms of degradation not included in the 1975–77 survey include deterioration in the quality of various ecosystems, such as forests and woodlands, flora and fauna, natural areas (scenery, wilderness, the 'national estate'), inland waters (both surface and groundwater), and the coastal zone (including estuaries and lagoons, mangroves and tidal flats, sand dunes, beaches, near-shore reefs and other marine habitats, but not discussed further here).

The pressures of clearing and agricultural practices on natural ecosystems range from subtle to profound and may have important consequences for both land quality and human well-being. Some of these consequences are discussed in this chapter (ecosystem damage, species loss, loss of genetic resources, threats from introduced pests), chapter 3 (damage from the use of agricultural chemicals) and chapter 5 (land management practices). Problems of water quality are considered under the broader implications of land degradation in chapter 4.

Clearing of forests and woodlands

According to the Department of Arts, Heritage and Environment (1986a), in 1788—when the first European settlers arrived in Australia—forests covered nearly 10 per cent of the country, and woodlands about 23 per cent. The more fertile or accessible areas were forested, and were also more attractive for agriculture. Consequently, it is not surprising that there has been a greater loss of forests than woodlands since that time. About 50 per cent of tall and medium height forests (30+ m and 10–30 m, respectively) and about 30 per cent of woodlands (where trees are smaller or more widely spaced) have been cleared or severely modified.

It should be noted that there is some disagreement between conservationists and foresters as to the extent of forest cleared since European settlement, with estimates ranging from more than two-thirds cleared to less than half. Some of this argument centres on problems of making estimates from inadequate historical data, and some from the problem of defining 'forest' (Cary and Barr, 1992). How dense and tall does woodland have to be, for example, before being reclassified? Steering a middle course, Graetz et al. (1992:113) agree with the above estimate that about half of Australia's forests have been cleared: from 690 000 km^2 in 1788 to 390 000 km^2 in the 1980s.

On a state-wide basis, the estimated proportional losses of forests and woodlands combined are shown in table 2.1. Victoria has the dubious distinction of having suffered the greatest proportionate loss of vegetation, followed by the ACT; the Northern Territory, not surprisingly, has lost very little. The greatest *absolute* losses occurred in New South Wales (a little over 300 million ha), followed by Queensland (about 300 million ha).

Table 2.2 provides data of the forest and woodland cover of the various states and territories in 1990. Data for plantations are also presented. Not surprisingly, eucalypts dominate. Rainforests cover only 0.28 per cent of Australia, and plantations (at 0.14%) are even less significant—a major concern if they are to reduce pressures by timber and paper pulp interests on our native forests.

Table 2.1 **Loss of forests and woodlands, expressed as a percentage of the areas covered by forests and woodlands at the time of European settlement.**

State	*Reduction of combined forest and woodland as % of pre-European cover*
New South Wales	50–55
Victoria	69–74
Queensland	35–37
Western Australia	31–43
South Australia	41–53
Tasmania	36–44
Northern Territory	0.03
Australian Capital Territory	60
Australia	36–40

Source: Wells et al., 1984; in Department of Arts, Heritage and Environment, 1986b: table 4.6.

Table 2.2 Areas ('000 hectares) of native forest and woodland in the Australian states and territories in 1990.

Type	*NSW*	*Vic.*	*Qld*	*SA*	*WA*
Rainforest	265	16	1237	–	–
Eucalypt	12761	5345	4795	–	2684
Trop. euc. & paperbark	–	–	4078	–	–
Cypress	1696	7	1686	–	–
Woodland	3300	2500	28200	900	20500
Plantation	298	227	192	99	104
Total	18320	8095	40188	999	23288

Type	*Tas.*	*NT*	*ACT*	*Aust.*
Rainforest	605	38	–	2161
Eucalypt	2327	–	51	27963
Trop. euc. & paperbark	–	2450	–	6528
Cypress	–	778	–	4167
Woodland	1051	7000	5	63456
Plantation	101	4	15	1040
Total	4084	10 270	71	10 5315

Source: compiled from ABS, 1992: tables 2.3.1 and 2.3.2.

Queensland has by far the largest area of rainforest (nearly 60% of the total which includes the temperate rainforests of Tasmania as well as tropical rainforests), while New South Wales dominates with its 12.8 million hectares of eucalypt forests, followed by Victoria and then Western Australia. At the other end of the scale, South Australia is distinguished by its total lack of forests, and woodlands are also poorly represented in that state. New South Wales has the largest area of plantation forests, but they cover barely 300 000 ha. The most rapid growth in plantations in recent years (the 1980–90 decade) occurred in Tasmania, where the area nearly doubled to 101 000 ha.

But there is a major problem with data here. The Australian Bureau of Statistics (ABS, 1992: table 2.3.1), for example, lists the total forested area in New South Wales as extending over 14.7 million ha (and 15.3 million ha in 1980); the Resource Assessment Commission's 1992 *Survey of Australia's Forest Resources* cites a similar figure (14.4 million ha). But the Department of Arts, Heritage and Environment (1986b), citing Wells et al. (1984), lists the forested area of New South Wales as covering between 32 and 62 million ha. Not only is the discrepancy between the sources very large, but the latter source explicitly indicates the inaccuracy of the estimates by giving a huge range between the minimum and maximum areas—although its minimum is still more than double the Bureau of Statistics figure. The *Australian Yearbook* provides similar figures as the Bureau of Statistics, and both quote their primary source of data as being 'state and territory forest services', as does the Resource Assessment Commission.

Although tree replanting around Australia accelerated over the last decade, particularly in response to the Greening Australia and Billion Trees programs (refer chapter 7), major net losses of native forests and woodlands are still occurring. Annual loss rates are not clear, but Victoria's *State of the Environment Report* (Scott, 1991) indicates that the average rate of tree loss in rural Victoria over the 25 years to 1983 was about 1 per cent each year. Between 1972 and 1987, more than 230 000 ha of native forests and woodlands were cleared, at a mean annual rate of 15 400 ha. Most of this occurred on freehold land, predominantly in western cropping regions (clearing restrictions on freehold land were imposed in 1991). Over the same period, reforestation of land already cleared by 1972 amounted to an annual average of only 4625 ha. If smaller tree plantings such as farm woodlots and shelterbelts are included, that figure rises to 8500 ha per year (Nadolny, 1991)—still considerably less than the clearing rate.

There have been huge rates of forest and woodland clearing in other states. In Queensland and New South Wales, average annual clearing rates (mostly of woodlands) of up to 500 000 ha were reported during the 1980s. In southwestern Australia, despite clearing bans in major river catchments and the prevention of further agricultural expansion into more marginal areas, approval was given to clear almost 130 000 ha in the three years to 1989 (Eckersley, 1989; Nadolny, 1991).

The data in the previous paragraphs refer to *areas* covered by various forms of vegetative communities; they provide no information concerning the *quality* or condition of the forests or woodlands. This question is addressed in the following section.

DEGRADATION OF WOODLANDS AND GRASSLANDS

Observations in the field, and a considerable amount of anecdotal evidence, suggest that the *quality* of Australia's vegetative cover has deteriorated along with the reduction in its spatial extent. But there seems to be little if any hard evidence to this effect. Botanists tend to focus on the status of *individual* species (discussed further below) rather than the condition of plant *communities* or ecosystems; and it is more than 20 years since Specht et al. (1974) reported on their 1971 survey of the conservation status of major plant communities (reassessed by Specht (1981) in 1979).

The 1971 survey (table 2.3) found that many plant communities were not represented in secure conservation reserves, and are therefore under

Table 2.3 Total number of major plant alliances per structural formation in each state of Australia, and the Northern Territory, and the percentages of these alliances not recorded or poorly represented in existing reserves.

Structural formation	*NSW*		*NT*		*Qld*		*SA*		*Tas.*		*Vic.*		*WA*		*Australia**	
	Total	*%*	*Total*	*%*	*Total*	*%*	*Total*	*%*	*Total*	*%*	*Total*	*%*	*Total*	*%*	*Total*	*%*
Tall closed forest/Closed forest/Low closed forest	14	43	2	–	23	52	–	–	26	19	6	50	8	63	79	39
Tall open forest/Tall woodland/Tall open woodland	5	20	–	–	2	–	–	–	18	11	11	45	5	40	41	24
Open forest/Low open forest	32	44	3	67	22	82	17	47	35	40	24	50	20	40	153	46
Woodland	23	43	7	29	19	95	8	25	12	33	11	55	22	50	102	52
Low woodland	26	69	17	53	14	93	13	31	19	68	6	33	18	39	113	58
Open woodland/Low open woodland	2	100	8	50	15	93	10	40	17	59	8	63	42	74	102	68
Closed scrub/Closed heath	12	58	–	–	8	75	2	50	39	13	10	40	15	47	86	35
Open scrub	16	38	2	100	5	80	19	37	14	57	9	44	12	42	77	45
Open heath	17	29	–	–	8	88	7	14	26	15	7	57	10	30	75	31
Tall shrubland	10	60	13	54	4	100	3	–	5	20	4	100	17	47	56	54
Low shrubland	7	43	6	100	7	100	18	39	11	18	4	100	5	40	58	53
Tall open shrubland	–	–	3	67	3	–	3	–	–	–	2	50	13	31	24	29
Low open shrubland	4	100	–	–	4	50	–	–	3	33	–	–	6	67	17	41
Hummock grassland	–	–	7	–	3	67	3	33	–	–	1	–	9	56	23	35
Grassland/Herbland (alpine and subalpine)	9	100	–	–	–	–	–	–	48	2	9	100	–	–	66	12
Grassland/Herbland} (lowland)	25	48	5	80	16	75	24	66	32	22	13	69	6	50	121	50
Total	202	45	73	53	153	78	127	40	305	25	125	57	208	50	1193	45

* As a few alliances occur in more than one state and are thus counted more than once, the totals for Australia will be greater than actual

} includes both open and closed structural forms

Source: Specht et al. (1974), from Department of Arts, Heritage and Environment, 1986b: table 5.1. © Commonwealth of Australia; used by permission.

threat of degradation if not already degraded as a result of human actions. The vegetation types which were least adequately represented were woodland and grassland communities, ranging from the brigalow woodlands of southern Queensland to the mallee of South Australia and Victoria and across the semi-arid grasslands to the Western Australian wheatbelt. According to Specht and his colleagues, the plant communities of the arid and semi-arid zones were particularly poorly conserved. Thirty-one forest and woodland alliances were not included in any reserves, and a further 15 were inadequately represented.

While there is little information on the condition of plant communities, the 1975 land degradation survey (discussed previously) provided some information on the condition of the vegetation of the arid lands (Department of Environment, Housing and Community Development, 1978). That survey found that more than 50 per cent of the vegetation in the arid zone was degraded to some extent, and that 13 per cent was severely degraded. The worst degradation was in South Australia and New South Wales, where saltbush and bluebush shrublands, semi-arid woodlands on the western fringe of the arid zone, and mulga shrublands had undergone the greatest change. On the other hand, spinifex grasslands, Mitchell grasslands, Mallee shrublands and most woodlands were in comparatively good condition. In areas that were extensively grazed, it was found that the species composition and plant density of the communities listed in the previous sentence were very similar to nearby experimental plots which had been protected from grazing for thirty years.

ECOLOGICAL DISTURBANCE

A good example of an ecological disturbance comes from the southwest of Western Australia. The jarrah leaf miner (*Perthida glyphopa*) has exploded in numbers to plague proportions over an extensive region covering both native forests and agricultural areas. The jarrah leaf miner is a moth, which lays its eggs on the jarrah (*Eucalyptus marginata*). The eggs hatch into larvae, which then eat circular holes in the leaves. In undisturbed environments, moth numbers are held in check by several species of native wasps, which parasitise the larvae (Wallace, 1970). But with extensive logging and clearing for agriculture, the habitats of the wasp have been severely reduced, such that there has been a population explosion of the moth. As a result trees covering extensive areas have been severely defoliated (about 400 000 ha of forest, and about a million ha of partly cleared land, with losses of between 64% and 83% of girth increment occurring in susceptible trees—Mazanec, 1974). In at least one recorded instance, the development of a saline seep has been attributed to the hydrological disruption caused by 'a severe attack of leaf miner' (Trotman, 1974:20).

Similar examples elsewhere in Western Australia, as well as other states, have been described by Hussey and Wallace (1993) and Davidson and Davidson (1992). Natural shrub, woodland and forest communities have very diverse flora and fauna species, but their clearing, the creation of pastures and drainage of wetlands have caused serious reductions in species numbers and diversity. For example, in Western Australia the loss of predatory birds (emu, ibis, butcher bird) has encouraged plague outbreaks of grasshoppers; and changes to the yate eucalypt environment have resulted in the loss of natural predators (birds, wasps), encouraging severe *psyllid* (a small, leaf-eating insect) attacks on the trees. Serious rural tree decline in other parts of Australia has also been attributed to the loss of predatory birds and increases in defoliating insects, and other pressures. Moreover, unprotected stands of remnant vegetation are often inadequate to support threatened species.

EXTINCT, RARE AND ENDANGERED SPECIES

Loss of species is one of the problems with which Australians are becoming increasingly aware and concerned. Vegetation loss, and alteration of vegetative structure, may result not only in the loss of plant species but also in the destruction or alteration of the habitat of fauna (animals and insects). Again, however, there is a difficulty with the data—not least because many Australian species, especially insects, have yet to be identified. If we do not know what we have, then it is difficult to assess what has been lost or is in danger of extinction.

According to the Department of Arts, Heritage and Environment (1986a,b), and the ABS (1992), the Australian fauna, while distinctive, are not particularly rich in terms of numbers of species (with the exception of insects). Vertebrate (backboned) fauna include about 300 mammals, 700 birds, 630 reptiles, 180 amphibians and 190 freshwater fish species. In contrast, about 70 000 insect species have been collected and described, and there is probably a similar number yet to be identified. Thousands of mollusc species are also undescribed.

Plant life is also rich. There are estimated to be 20 000 known species of vascular plants (flowering plants, ferns and conifers)—compared with about 4000 in Canada, a similar-sized country (Government of Canada, 1991)—together with many species of smaller plants, such as mosses and lichens, for which there are no reliable estimates.

Several species have become extinct since European settlement or are threatened with extinction. Indeed, it has been claimed that Australia's recent rates of plant extinction exceed those of any other continent (Kirkpatrick, 1991). However, Australia is also the most recently cleared continent: other regions may well have experienced higher, undocumented rates of extinction in the past.

From the information obtained by the Australian Bureau of Statistics (1992), 97 species of vascular plants have become extinct, and 3329 species are rare or threatened throughout Australia (table 2.4). The richness of Australian species is indicated by a comparison with Canada—with a similar history of settlement—which reports only 96 species of plants and animals which have become extinct, and only 550 threatened species (Government of Canada, 1991).

Western Australia has by far the largest number of 'presumed extinct' (70), endangered (91), vulnerable (363) and poorly known (573) species of plants of the Australian states—probably because of its greater species richness—although it should be noted that the local Department of Conservation and Land Management (CALM) provides much lower numbers in the latter three categories. Queensland, followed by New South Wales, has significant numbers of rare, poorly known and vulnerable species. In Victoria, 35 (and possibly as many as 100) species of native plants are now extinct, with 110 species endangered and 315 vulnerable. Less than 0.5 per cent of the original lowland grasslands of Victoria remain intact, whilst up to half of the natural wetlands have been destroyed or altered detrimentally (Scott, 1991). A detailed Australian list, including species names, degree of threat, conservation status and the states in which the species appear, is provided in Briggs and Leigh (1988).

The Department of Arts, Heritage and Environment (1986b) correctly points out that *Eucalyptus* is an especially important genus. They are almost totally restricted to Australia, are a major feature of many landscapes, especially in the more densely settled areas, and play a critical role in the life cycle of many plants and animals. Although no known species have become extinct since European settlement, 125 species (nearly a quarter of all known species) are at some degree of risk, especially in the east and southwest of Australia. Some of these communities are inadequately protected in reserves, particularly the woodlands of the southern half of the continent.

There are 20 species of mammals and 10 species of birds which have also become extinct, with greater numbers (including amphibians, reptiles and fish) endangered or vulnerable (table 2.5: some subspecies are also listed in the table). The Australian and New Zealand Environment

Table 2.4 Numbers of plant species by endangered status and state.

Status	*NSW*	*Vic.*	*Qld*	*SA*	*WA**	*Tas.*	*NT*	*Aust.*
Presumed extinct								
Known from type collection only	4	2	0	0	38	2	0	46
Geographic range <100km	5	1	4	0	23	3	0	36
Geographic range >100km	5	2	1	4	9	0	0	15
Endangered								
Known from type collection only	0	0	1	0	1 [0]	0	0	2
Geographic range <100km	34	6	24	10	71 [30]	6	3	148
Geographic range >100km	22	15	13	13	20 [4]	4	1	59
Vulnerable								
Known from type collection only	0	0	5	0	5 [0]	0	0	10
Geographic range <100km	87	17	102	21	252 [153]	19	7	495
Geographic range >100km	70	26	107	24	106 [45]	11	9	279
Rare (not threatened)								
Known from type collection only	0	0	0	0	0	0	3	3
Geographic range <100km	134	38	310	23	215 [105]	32	36	763
Geographic range >100km	145	59	273	57	129 [33]	51	50	601
Poorly known								
Known from type collection only	1	0	28	3	64 [70]	1	14	111
Geographic range <100km	7	2	116	6	255 [515]	3	11	398
Geographic range >100km	18	12	75	28	254 [427]	4	45	363

* Numbers in square brackets [ ] indicate Conservation and Land Management's (CALM) assessment when it differs from Rare or Threatened Australian Plants' (ROTAP) values.

Source: Briggs and Leigh (1988) in ABS, 1992: table 2.3.7.© Commonwealth of Australia; reproduced by permission.

Table 2.5 Extinct, endangered and vulnerable vertebrate species in Australia.

Amphibians	(7 endangered species, 2 vulnerable species)	
Endangered	Spotted Tree Frog	Sharpsnouted Day Frog
	Eungella Day Frog	White-bellied Frog
	Southern Day Frog	Eungella Gastric Brooding Frog
	Gastric Brooding Frog	
Vulnerable	Yellow-bellied Frog	Mount Baw Baw Frog
Birds	(10 species and 11 subspecies extinct, 1 species and 13 subspecies endangered, 15 species and 12 subspecies vulnerable)	
Extinct	Lord Howe Pigeon	Vinous-tinted Thrush
	Lord Howe Fantail	Norfolk Island Ground Dove
	Lord Howe/Norfolk Starling	Roper River Scrubrobin
	New Zealand Pigeon	Robust Whiteye/Silvereye
	Dwarf Emu	White Gallinule
	Paradise Parrot	Kangaroo Island Emu
	Lewin's Water Rail	Macquarie Island Parakeet
	Longtailed Triller	Macquarie Island rail
	Lord Howe Parakeet	Lord Howe Boobook
	Sth-Western Rufous Bristlebird	Norfolk Island Kaka
	Lord Howe Warbler	
Endangered	Malleefowl	Abbott's Booby
	Mount Lofty Southern Emuwren	Little Tern
	Coxen's Fig Parrot	Helmeted Honeyeater
	Gould's Petrel	Norfolk Island Parrot
	Blackeared Miner	Night Parrot
	Gouldian Finch	Regent Honeyeater
	Orange-bellied Parrot	Golden-shouldered Parrot
	Norfolk Island Thrush	Western Ground Parrot
	Western Partridge Pigeon	Western Long-billed Corella
	Sth-eastern Red-tailed Black Cockatoo	Norfolk Island Boobook Owl
		Melville Cicadabird
	Kimberley Crested Shriketit	Norfolk Island Silvereye
Vulnerable	Red Goshawk	Hooded Plover
	Carpentarian Grasswren	Noisy Scrubbird
	Eyre Peninsula Sthn Emu-wren	Redlored Whistler

Table 2.5 Continued.

	Lesser Noddy	Plainswanderer
	Christmas Frigatebird	Christmas Island Frigatebird
	Barrow Island Black-and-white Fairy Wren	Lord Howe Island Woodhen
		Blackbreasted Buttonquail
	Western Bristlebird	Christmas Island Hawk Owl
	Swift Parrot	Alexandra's Parrot
	Eastern Bristlebird	Western grasswren
	Eclectus Parrot	Southern Cassowary
	Abrolhos Painted Button-quail	Recherche Cape Barron Goose
	South-west Crested Shriketit	Dirk Hartog Black-and-white Fairy Wren
	Derby White-browed Robin	
	Western Grasswren	
Mammals	(20 species and 1 subspecies extinct, 26 species and 2 subspecies endangered, 17 species and 1 subspecies vulnerable)	
Extinct	Percy Island Flying Fox	Thylacine
	Christmas Island Rat	Central Harewallaby
	Lesser Bilby	Desert Bandicoot
	Eastern Harewallaby	Whitefooted Rabbitrat
	Alice Springs Mouse	Toolache Wallaby
	Lesser Sticknest Rat	Pigfooted Bandicoot
	Crescent Nailtail Wallaby	Shorttailed Hoppingmouse
	Gould's Mouse	Desert Ratkangaroo
	Longtailed Hoppingmouse	Darling Downs Hoppingmouse
	Broadfaced Potoroo	Bigeared Hoppingmouse
	Gilbert's Potoroo	
Endangered	Southern Right Whale	Kowari
	Burrowing Bettong	Blue Whale
	Western Quoll	Central Rockrat
	Humpback Whale	Dibbler
	Northern Hairynosed Wombat	Rufous Harewallaby
	Redtailed Phascogale	Heath Rat
	Banded Harewallaby	Julia Creek Dunnart
	Shark Bay Mouse	Barrow Island Euro
	Numbat	Longfooted Potoroo

Table 2.5 Continued.

	Brindled Nailtail Wallaby	Golden Bandicoot
	Northern Bettong	Western Ringtail Possum
	Greater Sticknest Rat	Leadbeater's Possum
	Brushtailed Bettong	Dusky Hoppingmouse
	Western Barred Bandicoot	Christmas Island Shrew
Vulnerable	Ghost Bat	Eastern Barred Bandicoot
	Sandhill Dunnart	Mountain Pigmypossum
	Greater Bilby	False Water Rat
	Blackfooted Rockwallaby	Goldenbacked Treerat
	Eastern Quoll	Brushtailed Rockwallaby
	Northern Hoppingmouse	Pilliga Mouse
	Proserpine Rockwallaby	Plain's Rat
	Western Mouse	Mulgara
	Pebble-mound Mouse	Barrow Island Euro
Fish	(7 species endangered, 6 species vulnerable)	
Endangered	Lake Eacham Rainbowfish	Brown Galaxias
	Trout Cod	Eastern Freshwater Cod
	Clarence Galaxias	Pedder Galaxias
	Swan Galaxias	
Vulnerable	Australian Grayling	Swamp Galaxias
	Saddled Galaxias	Ewens Pygmy Perch
	Yarra Pygmy Perch	Honey Blue-eye
Reptiles	(6 species endangered, 14 species and 1 subspecies vulnerable)	
Endangered	Western Swamp Tortoise	Loggerhead Turtle
	Pinktailed Legless Lizard	Broadheaded Snake
	Adelaide Bluetongued Lizard	
Vulnerable	Leathery Turtle	Yinnietharra Rockdragon
	Lancelin Island Striped Skink	Roughscaled Python
	Pedra Branka Skink	Airlie Island Ctenotus
	Striped Legless Lizard	Fitzroy River Tortoise
	Bronzebacked Legless Lizard	Legless Lizard
	Green Turtle	Pacific Ridley
	Hawksbill Turtle	Baudin Island Spiny-tailed Skink

Source: ABS, 1992: table 2.3.3. © Commonwealth of Australia; reproduced by permission.

and Conservation Council (ANZECC), which compiled table 2.5, lists a species as extinct 'if it has not been reliably detected for the last 50 years'.

It needs to be stressed that most species of plants and animals in Australia have suffered severe losses of extent or numbers, and diversity, since European settlement. Even abundant species of animals have lost particular populations and thus some degree of genetic diversity (Ride and Wilson, 1982). Pressures on the remaining communities and habitats are increasing, except in some of the better-managed reserves. While it may be true that more than 99.99 per cent of all species that have ever lived on this earth over geological time are now extinct (Margulis and Sagan, 1986:66–7, cited in Joseph, 1990:111), it is also true that the rate at which species are now being lost is unprecedented. 'The problem of disappearing species represents not only a conservation crisis on a scale never faced before, but also one of the greater environmental challenges we are ever likely to confront' (Kennedy, 1990:17).

LOSS OF GENETIC RESOURCES

Another less well recognised consequence of the clearing of natural vegetation is perhaps more significant at a global than a continental scale. This is the loss of a potentially useful pool of genetic resources—ironically, useful to agriculture as well as to medicine (Myers, 1980). In Australia, Benson (1990) has discussed the loss of native plants, while A.B. and J.W. Cribb (1981a, 1981b) have documented the wide range of useful native plants.

The loss of native plants is of concern because they may be more resilient, productive and nutritious than the bred varieties which replace them. Their loss has been particularly rapid in some regions, especially in south Asia. In India, for example, there were as many as 30000 varieties of rice in 1980; by 2000 it is estimated that as few as 12 varieties will dominate 75 per cent of that country (Tolba et al., 1992).

Modern agriculture's productivity has depended in part on specialised varieties of plants and animals often bred to certain conditions, but desirable qualities are not always permanent. The average lifetime of cereal varieties in Europe and North America, for example, ranges between five and fifteen years, and can only be restored by breeding programs. Furthermore, many of the major agricultural crops are bred around narrow genetic bases; for example, four wheat varieties produce 75 per cent of Canada's crop. In the continuing battle against pests and diseases, the diminution of a broad base of materials with qualities of vigour and resistance could threaten the long–term viability of major agricultural crops (Mooney, 1979; IUCN/WWF, 1980).

INTRODUCED SPECIES

People generally have become very aware of the problem of land degradation in recent years, particularly soil erosion and salinisation, and they are also aware that there is a degree of threat to Australia's indigenous plants and animals. There is some degree of awareness, too, of the fact that Australia plays host to some introduced species; but the problems these cause do not seem to have generated a great deal of interest or concern among conservationists. For this reason, an extended discussion is included here.

Several thousand pests, diseases and weeds adversely affect Australian agriculture; many are exotic in origin. There are also many introduced species which affect the natural environment. Their history of introduction is as early as the beginnings of European settlement (Bolton, 1981; Wace, 1988).

As summarised by the Department of Arts, Heritage and Environment (1986b), about 26 mammals, 24 birds, one reptile, one amphibian, 21 freshwater fishes and at least 2000 'foreign plants' have been introduced since European settlement and have now become established. In Victoria, nearly 25 per cent of all plants are non-indigenous species (Scott, 1991). There are also a number of introduced insects, other invertebrates such as snails, and disease-causing organisms. Some native species have also been introduced to parts of Australia in which they did not previously occur, such as the kookaburra to southwestern Australia. Here they are especially disruptive as predators among native nesting birds and small ground mammals and lizards.

Introduced plants and animals have had disastrous effects on many native species and have contributed directly to land degradation. Domestic and feral stock (domestic animals gone wild) feed on and trample plants and compete for water, while introduced carnivores, notably foxes and cats, prey on native animals which previously did not have similar predators. Huge areas in the arid and semi-arid zone in southern Australia have been severely modified as a result of overgrazing and burrowing by stock and rabbits. Feral goats, donkeys, horses and camels have become established even in the remotest parts of arid Australia. It has been estimated that in 1991 there were some one million feral goats in Western Australia: 814 000 were eradicated subsequently (Land Conservation Districts *Newsletter*, 21, 1993:5). Wetlands in the tropical north have been severely degraded by water buffaloes which feed, trample and wallow in them. Feral pigs have been responsible for similar damage in the north and east, and in forested areas. Introduced fish may have caused the decline of native fish, although many other factors have also been responsible for the deterioration of the waterways. Rats and mice feed on grains. Many feral animals are also important vectors for disease.

Weeds

In November 1992, an undated, draft National Weeds Strategy was distributed by the Plant Production Committee under the aegis of the Standing Committee on Agriculture, under the direction of the Commonwealth Department of Primary Industries and Energy (1992). 'Weed' is defined imprecisely, and in sexist terms, as 'any plant that is objectionable or interferes with the activities or welfare of man' (p. 9); but such a definition serves to make the point that 'weed' is a human concept. Another definition is to the effect that 'a weed is a plant which is out of place'. Although this definition is also vague, it is conceptually preferable in that it can be applied to natural situations and does not necessarily carry economic or utilitarian overtones. Nevertheless, it is the latter context which has led to a high level of concern.

According to the draft Weeds Strategy, of approximately 1900 plant species introduced into Australia since European settlement, nearly half are now weeds; and more than 220 of those have been proclaimed noxious weeds (in Canada, more than 500 plant species have become agricultural weeds (Government of Canada, 1991)). Nearly half of the noxious species in Australia were deliberately introduced as ornamental or agriculturally useful plants. Table 2.6 lists some of the major weeds by land-use categories; many have not yet reached their ecological limits of spread.

Examples of the problems posed by weeds include the following:

- Nearly 350 000 km^2 of northeastern Queensland have been invaded by rubber vine (*Cryptostegia grandiflora*). It is spreading along river courses and other moist habitats in semi-arid Queensland, and is having 'catastrophic' effects on the grazing industry. It creates impenetrable, smothering thickets, and the draft weeds strategy document refers to it as Australia's most important weed species.
- More than 80 000 ha of wetland in the Northern Territory have been infested by *Mimosa pigra*. The weed forms impenetrable thickets over vast areas and destroys floodplain habitats for birds, lizards and

Table 2.6 Some major weed species in Australia.

Land-use category	*Weed*
Rangelands	at least 20 native species, including turpentine (*Eremophilia sturtii*) budda (*E. mitchelli*) hopbushes (*Dodonea* spp.) punty bush (*Cassia eremophila*) brigalow (*Acacia harpophylla*) poplar box (*E. populnea*) EXOTICS INCLUDE: *Mimosa pigra* prickly acacia parkinsonia (*Parkinsonia aculeata*) mesquite (*Prosopis* spp.) rubber vine boxthorn (*Lycium ferocissimum*) *Acacia farnesiana* calatropis (*Calatropis procera*) chinee apple (*Zizyphus mauritania*) Athel pine (*Tamarix aphylla*) Noogoora burr (*Xanthium occidentale*)
Crops and sown pastures	annual ryegrass canary grass (*Phalarus minor*) paradoxa grass (*Phalaris paradoxa*) wild oats brome grasses (*Bromus* spp.) wild radish (*Raphanus raphanistrum*) skeleton weed fiddleneck (*Phacelia tanacetifolium*) bedstraw (*Gallium tricornatum*) *Bifora testiculata* Golden dodder (*Cuscuta campestris*)
Horticulture	blackberry nightshade (*Solanum nigrum*) spiny emex (*Emex australia*) caltrop (*Tribulus terrestris*)
Tropical crops	*Pennisetum pedicellatum* *Digitaria* spp. *Brachiaria* spp. sicklepod (*Cassia obtusifolia*) *Sida* spp. hyptis (*Hyptis* spp.) black pigweed (*Trianthema portulacastrum*)
Sown pastures	*Sida* spp. (NT) *Hyptis suaveolens* (NT) *partheneum* (Qld, NSW) blackberry (temperate areas) thistles (") serrated tussock (") *Vulpia* spp. (") *Sporobolus pyrimidalis* (coastal NSW & Qld) *S. africanus* (") Also, silverleaf nightshade (*Solanum elaeagnifolium*) field bindweed (*Convolvulus arvensis*)

Table 2.6 Continued.

Land-use category	*Weed*
	cutleaf mignonette (*Reseda lutea*) skeleton weed African lovegrass (*Eragrostis curvula*)
Aquatic and semi-aquatic systems	alligator weed water hyacinth (*Eichhornia crassipes*) salvinia (*Salvinia olesta*) *Lagarosiphon major* *Mimosa pigra* water lettuce (*Pistia stratiotes*) parrot's feather (*Myriophyllum aquaticum*) cabomba (*Cabomba carolinia*) paragrass (*Brachiaria mutica*) reed sweetgrass (*Glyceria maxima*) *Echinochloa polystachia* *Hymenachne amplexicaulis* Japanese kelp (*Undaria pinnatifida*)
Environmental weeds	Bitou bush/boneseed (*Chrysanthemoides monolifera*) bridal creeper (*Myrsiphyllum asparagoides*) boneseed pampas grass (*Cortaderia selloana*) rubber vine prickly acacia mission grass (*Pennisetum* spp.) mesquite *Mimosa pigra* blue thunbergia (*Thunbergia grandiflora*) salvinia water hyacinth Cootamundra wattle (*Acacia baileyana*) sweet pittosporum (*Pittosporum undulatum*) coast tea tree (*Leptospermum laevigatum*) umbrella tree (*Schlefflera actinophylla*)
Plantation forestry	improved pasture species blackberry woody weeds (e.g. wattles and eucalypts) St John's wort (*Hypericum perforatum*) naturally regenerated pine
Urban/industrial areas	ragweeds (*Ambrosia* spp.) pellitory annual ryegrass castor oil plant (*Ricinus communis*) oleander (*Oleander* spp.) rhus tree (*Toxicodendron succedaneum*) poison ivy (*Toxicodendron radicans*) *Grevillea* spp. Horsetail (*Equisetum arvense*) pampas grass scotch broom (*Cytisus scorparius*) St John's wort pampas grass blackberry

Source: Adapted from Department of Primary Industries and Energy, 1992: 13–18.

crocodiles. It has spread into Kakadu National Park and Arnhem Land, and may spread to Western Australia and Queensland.

- More than 50 million ha in western New South Wales and Queensland are affected by woody weeds, with some 500 000 ha being 'severely degraded'. It is considered that much of this area could become completely unproductive by the year 2020.
- Wild oats cost the wheat industry more than $40 million each year. This plant, together with three other common agricultural weeds—annual ryegrass, barley grass and capeweed—is now resistant to herbicides.
- The final examples concern the pollens from pellitory (which is spreading in metropolitan Sydney) and parthenium weeds. These pollens are a primary cause of hay fever and also trigger asthma attacks. Parthenium weed also causes dermatitis in people and animals. In Perth, South African veld grass is the curse of spring for urban dwellers.

While weeds are clearly a major problem for primary producers, they also pose a serious threat to the natural environment through displacing native plants and modifying animal habitats. They also threaten the health of people with allergy and asthma problems—although the draft strategy document fails to point out that this is also true of many pollinating native species. The weeds themselves and some of the methods used to eradicate them are thus both major types as well as causes of land degradation (chapters 3 and 4).

Other pests

Many other 'pests' have an adverse impact on the crop and livestock industries and constitute part of the degradation of those environments: many have been introduced. Major crop pests and diseases such as aphids, mites, nematodes, molluscs, fungi, viruses and bacteria can reduce production by as much as a third to one-half. Fungal diseases are devastating Western Australian and some Victorian native forests, woodland and shrub communities (Shearer and Tippett, 1989; Wills, 1993). Nearly $200 million are spent annually on fungicides and insecticides, mostly in agriculture (chapters 3 and 4), to control pests and diseases. Annual production losses to diseases in wheat alone amount to an estimated $400 million (Bureau of Rural Resources, 1989).

Parasites are a major problem of the livestock industry and incur an expenditure on veterinary chemicals of nearly $160 million a year to control them (Senate Select Committee, 1990). Internal parasites (especially intestinal worms) account for two-thirds of the costs; however, some Australian pests such as ticks are also a serious external parasite problem for stock. The wool industry, for example, incurs an annual cost of almost $500 million in terms of lost production and costs of chemicals to control internal and external parasites and blowflies (Bureau of Rural Resources, 1989). However, the bureau urges caution in interpreting the various costs which it considers are often overestimates, and which also vary considerably from region to region, year to year and industry to industry.

Some Australian native species are also viewed as 'pests'. Kangaroos, in particular, are believed to disadvantage the wool industry by more than $200 million in potential annual output as a result of grazing competition, compared with about $95 million attributed to rabbits (Bureau of Rural Resources, 1989). This has provided a potent basis for an argument by graziers to 'cull' kangaroos, but which has drawn serious opposition from conservationists and animal liberationists.

CONCLUSION

Chapters 1 and 2 have presented an overview of the main problems of rural land degradation in Australia. The overview has referred to accelerated soil erosion, deterioration of soil quality, decline of flora and fauna, and the spread of introduced species. Data limitations are a major problem, such that some material is both out of date and unreliable, and data on other environmental parameters, such as the quality of vegetation associations or plant communities, are simply not available or are subjective in nature. Nevertheless, it is clear that Australia has a major and growing problem.

Further environmental, social and economic implications of land degradation are discussed in the following chapters. Among the implications is the realisation that land degradation is not solely a problem for the rural land owner: its ramifications spread throughout the community. When it is recalled, for example, that the salinisation of water and soils in irrigated crop-growing areas is widely held to have been a significant factor in the fall of the Mesopotamian civilisation, it will be apparent that Australians need to take the problem seriously. Indeed, there are those who argue that following the end of the Cold War between the now defunct Soviet Union and the West, land degradation and its associated problems pose the greatest threat to the future of humankind.

■

3

Problems associated with the use of agricultural chemicals

There are several forms of land degradation in addition to those considered in chapters 1 and 2. One group is related to the ways in which farmers respond to the loss (or lack) of soil nutrients, and the spread of weeds and pests. Farmer responses commonly take the form of applying agricultural chemicals which, in turn, lead to a further array of environmental problems. This chapter considers, briefly, some of the more important of the adverse consequences associated with the increased use of synthetic fertilisers and pesticides.

ADVERSE EFFECTS OF SYNTHETIC FERTILISER APPLICATIONS

The big increase in fertiliser consumption in Australia since 1950 is apparent in table 3.1, which shows more than a six-fold increase of total amounts used. This growth rate is comparable to that exhibited by other developed countries and has been associated with significant increases in agricultural productivity (Tolba et al., 1992).

Pastures have been the major recipients of fertilisers, primarily of the phosphatic types which accounted for the leading fertilisers used in Australia until the mid-1980s, when nitrogen became the main fertiliser element used (table 3.1), especially on crops. Fertiliser application rates range from less than 100 t/ha/yr on pastures, through 100–300 t/ha/yr on

Table 3.1 Total Australian consumption of fertilisers, in Kt, in terms of elemental N, P and K, 1950–89.

	Nitrogen	*Phosphorus*	*K (Potassium)*	*Total*
1950	12	130	4	146
1955	20	186	13	219
1960	34	241	26	301
1965	69	370	47	486
1970	125	370	71	566
1975	175	315	80	570
1980	256	400	116	772
1985	350	340	115	805
1989	394	365	131	890

Source: Cribb, 1991: table 4.

cereals, to more than 1000 t/ha/yr on horticultural crops. It is the areas of high and continuous fertiliser applications which need particularly careful monitoring for some of the reasons described below.

The increasing extent and intensity of the use of synthetic fertilisers have resulted in a number of adverse environmental effects (Leece, 1974; McGarity and Storrier, 1986), which have also been well documented in other countries (OECD, 1986, 1989). In no particular sequence, the effects include water repellency, soil acidity, induced nutrient and trace element deficiencies, accumulation of heavy metal residues, reduction in numbers and diversity of soil organisms, eutrophication, and accumulation of nitrates in groundwaters.

Water repellency

One effect of the combined use of fertilisers and clover pastures has been an apparent increase in the water repellency of surface soil materials. An organic coating forms around the sand grains and repels water, especially when the soil is dry (McGhie and Posner, 1980). Water landing on these hydrophobic soils cannot infiltrate, and overland flow is virtually 100 per cent of incoming rainfall (less evaporation). The consequences of water repellency are a dramatic increase in erosion by overland flow, a lack of water in the soil, and delayed germination. The effect is often patchy, as can be seen in areas of bare soil surface in crops early in the growing season. The phenomenon is also seasonal. Once the early rains have been able to wet the soil, the hydrophobic effect is broken down.

Soil acidity

Increasing soil acidity is another major problem which seems to be associated with the use of fertilisers and leguminous pastures (Evans, 1991;

Scott, 1991). Soil acidity is also aggravated by removal of nutrients (in animals and crops), high stocking rates (through the addition of manure and urine to the soil) and high rainfall (which promotes leaching of the soil). On average, soil pH drops by one unit over 15–60 years, depending on the soil and how it is managed. Since about 75 per cent of Australia's pastures are less than 50 years old, there is a vast area likely to be affected by increased soil acidity over the next 50 years. About 24 million ha of agricultural land (about half of all agricultural land) are now affected by acidity in Australia, with more than 9.5 million ha of acid soils in New South Wales alone (Evans, 1991).

Once pH has dropped below about 5.0—as has already occurred in many of the older farming areas of Australia—a number of consequences become apparent. The nutrients phosphorus and molybdenum in the soil may become unavailable to plants. Aluminium and manganese become more mobile and may reach toxic concentrations, which can affect the growth of sensitive plants, causing structural damage and inhibiting nutrient uptake. The supply of the important nutrients boron, calcium, magnesium and potassium can be disturbed. Soil acidity can also affect the availability of soil nitrogen: nitrification by microbial processes can be reduced under acid conditions. With legumes such as subterranean clover, acidity affects not only nutrient availability but also nodulation. Increased acidity also allows the leaching of nitrates and with them, cations which are essential to plant growth.

Induced deficiencies

Imbalanced or heavy applications of fertilisers may lead to deficiencies or toxic excesses of major nutrients and trace elements. Plants may shrivel, distort, discolour or perform poorly. Excess nitrogen causes deficiencies of potassium, calcium or manganese; high levels of phosphorus cause deficiencies of zinc and potassium; high levels of potassium cause magnesium shortages, or chloride toxicities if potassium chloride is used; over-use of lime causes deficiencies of copper and iron; and high applications of superphosphate can reduce the availability of selenium and copper (the latter if molybdenum is in the soil or added in the superphosphate) which has important implications for animal health.

Heavy metals

There are many impurities in synthetic fertilisers—such as fluoride, lead, cadmium, zinc and uranium—and these may exert a range of adverse environmental and health effects (Tiller, 1989; Fergusson, 1990). Heavy metals behave in different ways depending on the plant species, soil pH, organic matter content, soil moisture, cation exchange capacity and sesquioxide (Fe_20_3 and Al_20_3) content. Generally, heavy metals are strongly adsorbed on to organic matter.

Cadmium is the best studied of the heavy metals, and its contamination of agricultural soils poses something of a problem in Australia. On average, about 40 parts per million (ppm) of cadmium contaminated superphosphate in Australia (CSIRO, 1975). Concentrations have increased by 0.1 to 0.2 ppm in most farm soils receiving normal rates of superphosphate application over the last 50–60 years. Concentrations are greater on 'heavier' (higher clay content) or acid soils. Market garden soils have yielded cadmium concentrations 10 times higher than in adjacent, uncultivated soils (the use of sewage sludge on these soils is also a major source of cadmium).

Cadmium and other heavy metals can inhibit water uptake, photosynthesis and growth in plants and also adversely affect soil biological activity. Vegetables, especially root and leafy crops, seem to concentrate heavy metals more readily than pastures and grain crops, and this has important implications for human health (Tiller, 1989).

Heavy metal poisoning can cause: kidney disease; decalcification of the skeleton; excess protein, calcium and amino acid excretion; tiredness; shortness of breath, and impaired smell in humans. Cadmium is the main element of concern and it has exceeded permissible concentrations in foodstuffs in Australia. Continued monitoring of the role of cadmium in agriculture has been recommended (Simpson and Curnow, 1988).

Soil organisms

A range of macro, meso and micro soil organisms can be affected by the heavy or prolonged use of fertilisers, which can reduce population numbers or shift species composition. Earthworms, for example, are killed by direct contact with phosphatic fertilisers. On the other hand, an indirect benefit for worms and other soil organisms may result from increased pasture growth and soil organic matter content which increases their food supply. The result of reduced soil organisms may be to affect the availability of soil nitrogen, the production of ethylene gas, the regulation of soil microbial activity, the supply and regulation of plant nutrients, and the incidence of soil-borne plant diseases (Marshall, 1977).

Eutrophication

Agricultural fertilisers can be transported away from their sites of application through volatilisation, leaching or via eroded soil particles. Plants commonly use between 30 per cent and 70 per cent, and as little as 10 per cent, of fertilisers applied to the soil, leaving the remainder elsewhere in the environment (McGarity and Storrier, 1986). These 'recovery rates' are affected by the type of fertiliser, frequency of its use, soil and crop type, ambient conditions and management practices. Some of the consequences of the translocation of fertilisers and their residues away from the farm are discussed in this and the following section.

Eutrophication is a consequence of excess quantities of nutrients—usually nitrogen and/or phosphorous—entering water bodies (streams, rivers, dams, lakes, estuaries). The major sources of contamination result from direct and indirect runoff and discharges from both urban and rural land, from sewerage systems, abattoirs, intensive animal industries, fertilisers, sediments and organic materials. Algae which feed on the nutrients 'bloom' and then die, consuming dissolved oxygen in the water. The resulting lack of oxygen kills fish, and some of the algae, particularly blue-green species, are themselves toxic. In addition, water quality (taste and odour) is reduced, costs of treatment increased and recreation value reduced. Transport and drainage may be impeded and flooding increased. Eutrophication is a problem throughout Australia, as it is exacerbated by the ephemeral or seasonal nature of many streams and rivers (producing low flows or shallow water bodies), and high temperatures.

Many areas are affected by eutrophication. Some examples around Australia include the Peel Inlet (Western Australia), Mt Bold (South Australia), the Gippsland Lakes (Victoria), Hay Weir Pool and Carcoar Reservoirs (New South Wales) and North Pipe dam (Queensland) (Cullen, 1991). There have also been reports of 'greening' in Lake Argyle in the far north of Western Australia (Costin, 1991). Surveys of Victorian surface waters have shown that a high percentage of monitored sites, especially in heavily cleared agricultural catchments, experience 'poor to degraded' conditions with respect to nitrogen and phosphorus concentrations, implying a potential for eutrophication to occur (Scott, 1991). Indeed, such problems have been especially severe, and intensifying, across the Murray/Darling basin; and in 1991/92, Australia experienced its worst outbreak of toxic blue-green algae in the Murray/Darling system as well as in some coastal rivers (figure 3.1).

According to Cribb (1992), a report on the outbreak produced for the Murray/Darling Basin Commission by Gutteridge Haskins and Davey (1992), found that 'diffuse' pollution sources—nutrient and sediment runoff from grazing, dryland farming and forests—comprise the largest general source of nutrients entering the river systems, with phosphorus

Figure 3.1 Occurrence of blue-green algal blooms in New South Wales during 1991/92 (not including farm dams).

Source: Based on Higgins, 1992.

loads averaging 950 tonnes a year. Irrigation farming was responsible for between a third and a half as much pollution in the Murray/Darling system as that from urban sewage.

The biggest point-source contributors of nutrients are the sewage works at Albury–Wodonga and Shepparton, the Sunraysia tile drains and the reclaimed swamps of the lower Murray in South Australia. On the Murrumbidgee, Canberra's Lower Molonglo sewage treatment plant contributes the most nitrogen and Wagga Wagga the most phosphorus. In the Darling system, Toowoomba, on the Dividing Range in Queensland, is the largest point-source contributor. Victoria's Shepparton region and the lower Murray in South Australia contribute the highest nutrient inflows from irrigation; while the significance of pollution from cattle feedlots, piggeries and fish farms, and runoff from dryland farms, needs more investigation.

Nitrates

Nitrogenous fertilisers are often implicated in nitrate (NO_3) pollution of ground and surface waters. Natural concentrations of nitrates in Australian waters are generally low, at around 5–10 mg/l (compared with the World Health Organisation's (WHO) standard for drinking water of 50 mg/l).

However, a national survey of groundwater in Australia found that nitrate concentrations exceeded 20 mg/l at more than 1200 sites, at depths of up to 150 m (Lawrence, 1983). This contamination was usually associated with human activity, with the highest concentrations being found in areas of intensive cultivation (horticultural and orchard areas in New South Wales, Victoria, southwestern Western Australia and southeastern Queensland). Nitrogenous fertilisers and soil mineralisation from cultivation were the major contributors. Other problem spots were related to the use of septic tanks in unsewered areas of Perth and in Victoria, where discharges from soak wells have nitrate concentrations of around 100 mg/l; dairy processors (in South Australia and Victoria); piggeries, and industrial and mining activities. More recently, a survey revealed nitrate concentrations exceeding drinking water guidelines over an area of 1000 km^2 in the Mt Gambier region of South Australia. It was considered that clover pastures and grazing animals, rather than dairy processors and abattoirs, were the main contributors (Scott, 1991).

These higher readings equate with findings from a number of North American and European studies. In southern England, for example, 20 per cent of water samples breach the European Community's 50 mg/l drinking water limit for nitrates, with a number exceeding 100 mg/l (Burt et al., 1993). And Black (1993) has reported that 83 sites in California had nitrate concentrations in water ranging from 6 to 222 mg/l. In Germany, nitrates have been reported at depths of 10 m, with the highest concentrations being found under intensely fertilised vegetable crops.

Of greater concern is that nitrates, which have been leaching from soils over 20–30 years, are moving into groundwaters around Australia, associated with a huge (ten-fold) increase in the use of nitrogenous fertilisers (from 34 Kt in 1960 to almost 400 Kt in 1989: see table 3.1). This suggests the possibility of a serious problem of nitrate pollution developing over the next two decades.

Elevated nitrate concentrations in food and water have several potentially adverse health consequences (with health problems in adults generally occurring at concentrations of 100+ mg/l), although the severity of these risks has been disputed (Burt et al., 1993; Black, 1993). Nitrates convert into nitrites (NO_2) in the body and at excessive concentrations affect the body's ability to transport oxygen, causing cyanosis and, in serious cases, death. Children under the age of one, and young animals, are especially susceptible to nitrate poisoning from drinking water with concentrations exceeding 10 mg/l, and may experience a condition known as methaemoglobinaemia ('blue baby syndrome'). Other adverse effects in humans and animals include thyroid dysfunction, vitamin A deficiency, impaired cardiac function, liver damage, and increased abortion rates. Nitrates may also combine with amines and amides present in food and form nitrosamines, which are known to be potent carcinogens. Black (1993) notes that the average daily intake of nitrates in food is about 17 mg, mostly from vegetables, and that this is a more important source than polluted waters.

Epidemiological evidence on nitrate-induced cancers is conflicting, although animals are more susceptible than humans due to the nature of their digestive systems. O'Riordan (in Burt et al., 1993), in observing the various controversies surrounding nitrates, advises that the 'precautionary principle' (when in doubt concerning the consequences of a proposed action, either do not act or, at the very least, proceed with the utmost caution) should apply in management and policy making in relation to nitrates. This is particularly appropriate in that nitrate concentrations will continue to increase, and given that there may be a long time lag between exposure to nitrates and health outcomes.

ADVERSE EFFECTS OF USING SYNTHETIC PESTICIDES

Another consequence of the replacement of natural ecosystems with introduced agricultural crops and pastures is the emergence of weed, insect and disease problems. Ecological disruptions disturb or remove the habitats of natural predators, which may be destroyed or reduced severely in numbers and extent. Population imbalances of pest species result from such disruptions (Pimentel and Edwards, 1982).

As with the problem of declining soil fertility, modern agriculture tackles growing weed, pest and disease problems with synthetic chemicals. In both instances it can be argued that it is the symptoms of the problems which are being tackled, not the causes. And, as is the case with synthetic fertilisers, the increased application of pesticides has given rise to a complex array of further problems. In essence, the difficulty is that pesticides by their very nature are biocides, designed to kill or disrupt organisms, and as such have the capacity or potential to harm all living things, some more directly than others. For the most part, biocides and their breakdown products are mobile in the physical environment, and residues and metabolites may directly or indirectly affect plants, animals and humans. Reviews of the behaviour of pesticides in the environment include Guenzi et al. (1974), Brown (1978) and Schnoor (1992).

Toxicity of pesticides

The toxicity of a chemical has an important bearing on its impact on individuals, populations and the environment. While acute effects are directly obvious, sub-lethal or long-term effects can seriously affect individual or system resilience.

The toxicity of pesticides is measured in several ways. Acute toxicity is evaluated in terms of oral (ingested) or dermal (absorbed through the skin) Lethal Doses (LD), doses which are determined by the concentration of active ingredient in milligrams per kilogram required to kill 50 per cent of test animals in a laboratory (LD50). The lower the figure, the more toxic the chemical. For example, the highly toxic chemical aldicarb has an oral LD50 of 0.93 (and a dermal LD50 of 5), while the oral LD50 of the mild herbicide atrazine is 2000 (and the dermal LD50 is 7500) (DPI, 1980).

Other measures of toxicity, under controlled conditions, involve observation of the effects of short, long-term and chronic dosages on test animals, and the influence of the chemicals as mutagens (cell damaging), teratogens (foetal damaging) and carcinogens (cancer forming).

However, there are important reservations concerning toxicity data. They are obtained by controlled testing in laboratories using small, purpose-bred animal populations. Testing generally does not take into account factors such as differences in age, race, sex, diet, immune deficiencies, genetic weaknesses or the effects of stress. Chronic feeding studies observe the outcome of a specified period of feeding test animals with doses of a pesticide. But acute and chronic toxicity studies ignore the sub-clinical or delayed effects of once-off or chronic exposure to one or more pesticides. Such effects may include impaired immunity, respiratory, circulatory, behavioural or reproductive disorders, allergies, general health impairment, cancer, genetic or birth defects.

A particular problem concerns the possible synergistic effects of one chemical interacting with thousands of other chemicals and organisms with which mammals come into daily contact. Worldwide, there are an estimated 60 000 commercially produced chemicals in common use, mostly organic chemicals and pesticides, with about 1000–1500 new ones introduced each year (OECD, 1981). While not all of these are necessarily 'harmful', a Chemical Load Crisis conference held in Sydney criticised health regulations for ignoring the harmful effects of combinations of toxic chemicals which compromise the human immune system (*The West Australian*, 24 July 1993:41). Added to this are major research difficulties in identifying cause-effect relationships between exposure to a chemical or chemicals and health outcomes, in having to extrapolate from laboratory animal studies to humans, and in relying on inadequately designed or interpreted epidemiological studies.

Pesticide and veterinary chemical use in Australia

There has been a big increase in the use of agricultural pesticides in Australia in recent years (figure 3.2), with a particularly marked growth in herbicide use. Overall, the crop industries are the major pesticide users (see table 4.4). There are no accurate data available on the total quantity of pesticides used in Australia. However, extrapolating from estimates made in 1986 (Conacher and Conacher, 1986) and the work of others (Swarbrick, 1986; Pimentel and Levitan, 1986; Mowbray, 1988), we estimate that in 1993 at least 30 000 tonnes of pesticides were used in Australian agriculture.

Veterinary chemicals

Another major input of chemicals on farms (mainly livestock operations: see table 4.4) is veterinary chemicals (figure 3.2). The control of parasites is the leading category of use, accounting for over two-thirds of expenditure on all veterinary chemicals in 1988 (Senate Select Committee, 1990:117).

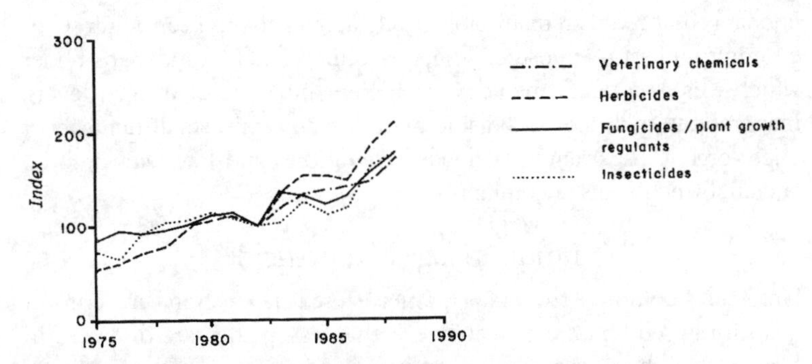

Figure 3.2 Agricultural and veterinary chemicals, index of quantities sold in real terms in Australia 1975–88.

Source: Bureau of Rural Resources submission to the Senate Select Committee Inquiry into Agricultural and Veterinary Chemicals, Hansard, 19 July 1989:2871.

The main sources of environmental concern with respect to veterinary chemicals are the contamination of livestock products (hormones, dips, antibiotics) and the residues in soil and water (from animal wastes, improper use and disposal of the chemicals). There is also evidence that serious resistances to anthelmintics (chemicals for treating intestinal worms) have developed in livestock, and that antibiotics indirectly may lead to bacterial resistance problems in humans. Certain livestock hormone treatments may cause hormone disturbances in humans.

Pesticides in the environment

The major concern over pesticide use in agriculture is that the physical environment ultimately becomes the sink for these chemicals (along with others) and their breakdown products.

The degree to which a pesticide poses an environmental risk is determined by the rate and frequency of its use and its persistence, mobility, ability to bioaccumulate (to concentrate in living tissues, especially animal fats), and toxicity to non-target organisms. Under ideal conditions, and because of their organic structure, many modern pesticides are expected to degrade readily in the environment from the effects of temperature, light, moisture, oxygen, organic matter and micro-organisms, experiencing what is known chemically as first-order kinetic reactions. However, the half lives of pesticides vary from hours, to days or months. Some, such as the organochlorines and certain herbicides, are designed to persist. Organochlorines (now largely discontinued) accumulate readily in food chains.

But other pesticides that would 'normally' be expected to degrade quite rapidly can also experience long periods of being relatively unchanged and thus remain biologically active, giving rise to unintended environmental exposure. The variability of field conditions under which this may occur includes insufficient moisture or oxygen levels, inappropriate pH, lack of organic matter, sandy soils, small numbers of micro-organisms, swampy or flooded conditions, and low temperatures, and are by no means uncommon.

Minimum tillage and pesticides

Herbicides comprise the major pesticide used in Australia in terms of quantities used and areas treated. For this reason they are discussed in greater detail here, recognising however that fungicides and insecticides (especially in intensive horticulture, cotton and sugar cane industries) are not without their impacts.

Early in the 1970s, the world experienced an 'oil shock', when oil prices increased rapidly to unprecedented levels. For farmers, the cost of fuel became their most expensive cost item after interest rates (figure 3.3). Thus many looked for less expensive means of preparing their land for seeding than the conventional cultivation practices. In this respect, minimum tillage was a popular choice. This term encompasses several tillage techniques which range from no tillage and direct tillage (or direct drilling) to reduced tillage or minimal cultivation. Differences between the methods depends on the number of cultivations used (in all cases, fewer and shallower than normal cultivation practices), the types of tillage machinery used and the type and frequency of herbicides used.

In the wake of the oil crisis during the 1970s, several global companies were actively seeking to expand their sales of herbicides. In Australia, the Western Australian wheatbelt was targeted by a British company, Imperial Chemical Industries (ICI), which at that time was marketing a pre-plant, knock-down herbicide known as Spray.Seed (a paraquat/diquat mix). In association with direct drilling techniques, the first field trials with Spray.Seed took place in 1971 on 10 000 hectares. The area treated increased dramatically to approximately two million hectares in 1984 (by then including other herbicide products) (figure 3.4). But in addition to the growth in direct drilling, areas treated with other herbicides and other minimum tillage methods had also grown rapidly in the early 1980s: by 1986, it was estimated that a total of 4.3 million hectares were being treated with herbicides (ABS, 1986).

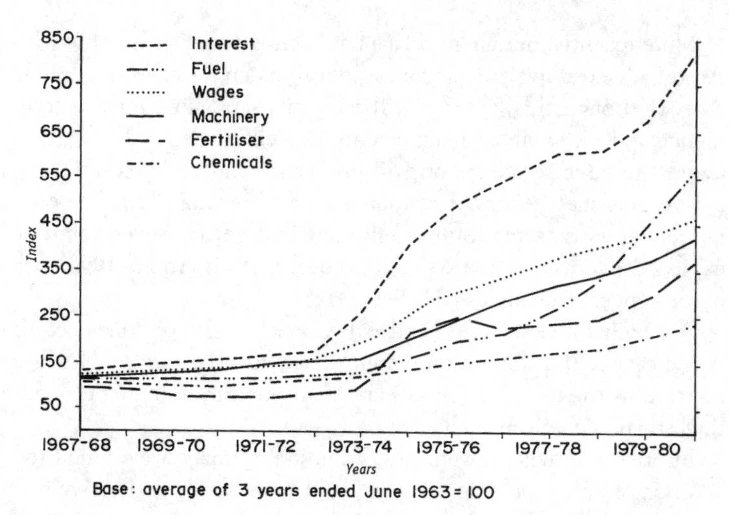

Figure 3.3 Prices paid for selected farm inputs in Australia 1967/68–1979/80.
Source: J.L. and A.J. Conacher, 1986:70.

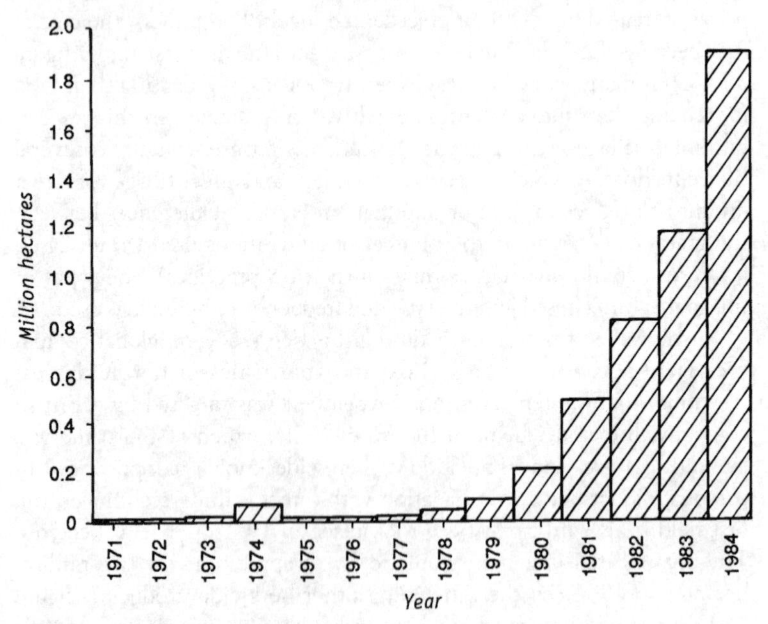

Figure 3.4 Growth in the area of crops treated with direct drill herbicide in Western Australia 1971–84.
Source: J.L. and A.J. Conacher, 1986:33.

More recently, minimum tillage has been adopted widely throughout Australia's extensive cropping areas, not so much in order to reduce fuel costs (although this factor is still relevant), but more for the convenience and lower labour requirements, as well as the need to reduce or avoid the adverse effects on soil properties caused by conventional tillage practices. About three-quarters of Australia's broadscale crop farmers now use some form of minimum tillage. However, many still express concerns over the system (Gorddard, 1991; Anon., 1993); some of the concerns are discussed below.

A detailed review by J.L. and A.J. Conacher (1986) concluded that the effects of the shift to minimum tillage and the associated massive increase in the use of herbicides are often unclear, a finding echoed in English and American studies.

In brief, the major advantages of minimum tillage were found to be agronomic and economic. Weeds could be controlled effectively and conveniently on a large scale with a range of herbicides, resulting in substantial reductions in time, labour, machinery and fuel costs. Additional advantages included positive changes to soil physical, chemical and biological properties, and reductions in erosion rates, often translating into improved crop yields. These changes were especially important where agricultural soils had been degraded by conventional methods of cultivation, and productivity impaired. The system also enabled farmers to develop otherwise unsuitable marginal lands for cropping and to close up cropping sequences. However, it is important to note that the responses to minimum tillage have varied widely across different soil types, tillage methods and regions.

With respect to adverse effects on the environment and humans, the research literature indicates that, in general, agricultural herbicides (especially in comparison with some insecticides and fungicides) do not cause any major disruptions if applied carefully at recommended rates. However, the current absence of any conspicuous problems associated with these chemicals may reflect the lack of research focus, inadequate technologies for measuring trace residues and their effects, or absence of precedence. It is not uncommon to find Australian scientists making strong pleas in the literature for more research into areas such as the effects of herbicides on total ecosystems, soil organisms, groundwaters, food and humans (for example, see the recommendations in Senate Select Committee, 1990).

Examples are provided by the Conachers (1986) of the direct effects of herbicide use in the Australian environment. The problems include: damage to sensitive crop species; a shift from broadleaf to grass weed dominance; the emergence of herbicide-tolerant and resistant weed species; amplified pest and disease problems such as take-all, root rot, nematodes, spider mite, lucerne flea and webworm (which have resulted in the increased use of insecticides and fungicides); herbicide residues which damage subsequent crops or pollute surface and ground waters; loss of restorative pasture and especially legume phases, and the increased need to use nitrogenous fertilisers.

Less easy to estimate are the indirect effects such as the 'cost' of drift damage to natural ecosystems. Herbicide contamination of surface and ground waters or food crops is also of direct economic importance if, for example, residues are responsible for damage to irrigated crops, or if food and water are rendered unfit for human or animal consumption. These types of damage have occurred, but few attempts have been made to measure the costs. The total of the direct and indirect costs could negate the 'expected' (and debatable) economic return on herbicide investment (see chapter 4).

In conclusion, there are both reasonably clear benefits and costs associated with minimum tillage and herbicides. There is also a vast, grey area of uncertainty. Given this uncertainty, the enormous and rapid expansion of herbicide use constitutes grounds for concern. If we are to err, it should be on the side of caution: in other words, the precautionary principle should apply.

Pest resistance to pesticides

The problem of increasing pest resistance to pesticides is a global one, with over 500 pest species affected (Tolba et al., 1992). In Australia, serious resistances to a range of chemicals have been experienced, in the orchard (spider mite), cotton (*Heliothis* moth) and livestock (intestinal worms) industries (for further details refer the Senate Select Committee, 1990 and the submissions to it, recorded in Hansard).

Such problems cause direct costs in terms of additional chemicals needed and lost production, and indirectly, the costs in terms of damage to the environment. One dramatic example of pest resistance comes from the failed Ord River cotton venture. In its last year of operation, it was necessary to use over 40 spray rounds of DDT per season, up to 125 kg/ha ('normal' use would be less than 10 kg/ha). High DDT residues were found in local aquatic organisms (CSIRO, 1980; Woods, 1981). Similar problems were experienced in the Namoi cotton area (State Pollution Control Commission, 1980). But even the use of less toxic synthetic pyrethroids has resulted in the development of pest resistances in the present-day cotton industry.

It is therefore particularly unfortunate that increasingly favoured, non-chemical alternatives are showing unexpected signs of failure. One example is the development of resistance to the biological control agent *Bacillus thuringiensis*, thereby placing the development of transgenic crops (plants with introduced genes) under question (Holmes, 1993).

The need to use broader-based systems of pest management becomes apparent.

Spray drift

Rarely does the total amount of chemical applied reach its target, and one estimate has suggested that in some cases as little as 0.1 per cent may reach the target pest (Pimentel and Levitan, 1986). The potential for pesticides to drift or volatilise can be high, especially if appropriate application conditions are not observed. As the Conachers (1986) demonstrated, there have been many examples of non-target damage from herbicide spraying in Australia, with consequences including damage to native vegetation, other crops, bees and wildlife, as well as contamination of soil and water.

The direct costs can be high in terms of lost income or compensation claims (several exceeded $30 000 in Western Australia in the early 1980s). Although difficult to measure, damage to ecosystems may be profound in terms of loss of structure, function, productivity and stability (Pimentel and Edwards, 1982).

The contamination of surface- and groundwaters may cause serious difficulties. There may be long lag times before the extent of contamination is fully known, it can be difficult to remove (or have access to) the contaminants, synergistic effects with other chemicals are possible, and there are potential risks to human health. There is substantial evidence that *normal* agricultural use of pesticides has polluted shallow aquifers in Europe and in 26 states of the United States (OECD, 1986; NRC, 1989). Over 80 per cent of public drinking water supplies from surface water sources in Iowa and Ohio tested in the mid-1980s had two or more pesticides present. Examples are also known from Australia (Scott, 1991). The problems are especially marked in areas of vulnerable hydrogeology, intensive agricultural use, or as a result of improper use and careless disposal of containers.

Residues in food

Australian authorities regularly monitor domestic and export produce for residues in food. In general, violation levels do not exceed 2 per cent (Senate Select Committee, 1990), although there is room for improvement in the extent and regulatory nature of monitoring, as recommended by the Senate Inquiry.

Higher violations are generally recorded in the fruit and vegetable groups than for other foods, and there continue to be relatively high concentrations of organochlorines in a range of animal foodstuffs. Increased controls over the agricultural use of persistent organochlorines have been reflected in declining levels in food. However, non-agricultural sources such as termite control, illegal or past uses (on orchards, cotton, tobacco and potatoes), continue to pose problems with substances such as dieldrin and chlordane. For example, the most recent human breast milk survey in Victoria revealed that most samples still contained organochlorine residues.

Although the means for DDT and HCBs had declined over the period 1970–86, dieldrin concentrations in breast milk remained essentially unchanged. The daily intake for babies exceeded WHO/FAO standards for HCB, DDT, PCBs and dieldrin in 32 per cent, 42 per cent, 50 per cent and 97 per cent of the samples tested, respectively. However, health authorities concluded that the benefits of breastfeeding still outweighed the disadvantages of chemical contamination (Luke et al., 1988).

Pesticide residues and human health

The effects of trace amounts (via food, water and air) of pesticide residues on human health are difficult to determine and can be expected to vary with the subject's age, sex and general state of health. Nevertheless, it has been demonstrated that chronic exposure may cause behavioural changes, blood disorders and vital organ damage, and that some pesticides act as mutagens, carcinogens and teratogens (Pimentel et al., 1980; Kaloyanova, 1983; Dosman and Cockroft, 1989). The list in table 3.2 of 'properties and criteria of concern' with reference to some agricultural pesticides from the Victorian *State of the Environment Report* (Scott, 1991) is instructive.

Much of the poisoning due to acute and direct exposure arises from improper use. Lack of care in mixing and using chemicals, spray drift, careless application around water bodies, incorrect disposal of washdown wastes and pesticide containers, all contribute to unwanted contamination of soils, water and atmosphere. Some of the health consequences are illustrated by the following examples.

In 1991, the Victorian Department of Agriculture received 109 complaints concerning spray drift. Sixty-one of the complaints were in relation to ground applications (the remainder were applied from the air), and they were more or less evenly divided between concerns over damage to crops or livestock and effects on human health (Scott, 1991: table 24.8). Western Australian surveys showed that in 1985, 35 per cent of farm chemical users reported ill effects from crop spraying. In 1993, of 920 farmers surveyed nationally, 36 per cent of chemical users experienced ill effects from chemicals. Over half (59%) of these were attributed to herbicide use (figure 3.5), with the symptoms reported being illustrated in figure 3.6. Implementation of safety precautions (such as wearing overalls, gloves, respirators and plastic boots) could avoid a large proportion of these symptoms.

With respect to acute poisonings, data for Australia are poor as there is no national system for reporting. However, a report in 1983 from the now defunct Commonwealth Department of Health Poison Centre recorded over 5000 poisonings due to pesticides, representing 8 per cent of all chemically related poisonings in Australia. Although a high proportion were accidental poisonings of small children, it is likely that many cases of occupational, pesticide poisoning were not recorded (being either misdiagnosed or not considered serious enough to seek medical help).

According to the World Health Organisation, 60–70 per cent of all pesticide poisonings are of an occupational nature (WHO, 1990).

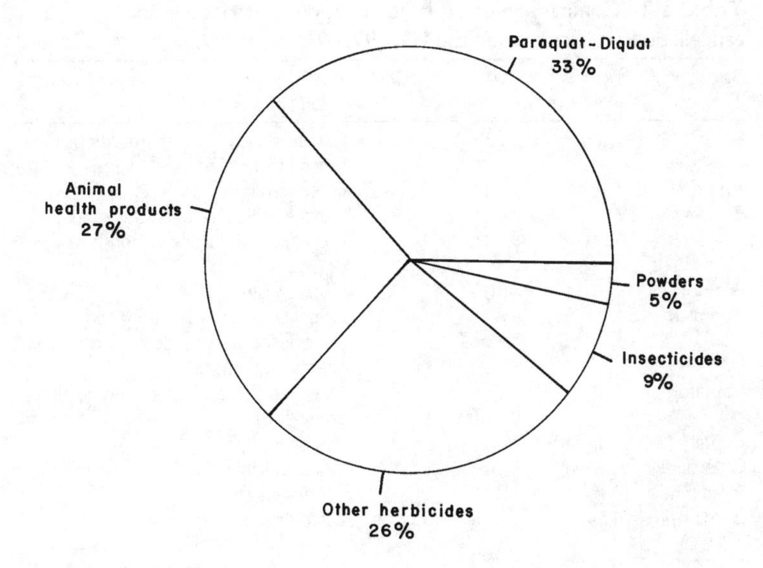

Figure 3.5 Farm chemicals responsible for ill health, from the Kondinin National Agricultural Survey held in April 1993.

Source: Kondinin Group, 1993:28.

Detailed studies on occupational poisonings by pesticides in 1983 summarised by the California Department of Food and Agriculture (1984), indicated that the majority originated in agricultural industries. Thus, current levels of pesticide poisonings in Australia could be much higher than 5000 per year, representing a significant cost in terms of health care, time off work and compensation payments.

By virtue of their more acutely toxic nature, insecticides generally account for the largest proportion of pesticide poisonings, and commonly affect the central nervous system. In contrast, a number of fungicides (not all—some are very toxic) and most herbicides tend to cause less severe symptoms—such as skin and eye complaints. Suicides figure prominently among the fatalities.

Conclusion

Land degradation is a very complex issue. It is not merely a problem of accelerated erosion by water or wind, or the consequences of clearing too much bush. This chapter has shown that trying to deal with problems of declining soil fertility, and the spread of weeds and insect pests,

Table 3.2 Comparison of available pesticides, restrictions and cancellations in Victoria and the US, 1990–91.

Agricultural chemical	*Victorian registration*	*US registration*	*Properties and criteria of concern*[a]
Aldrin	C (1989)	C[b]	Carcinogenicity, bioaccumulation, hazard to wildlife
Amitraz	A	R (L)	Oncogenicity
Benomyl	A	A	Mutagenicity, teratogenicity, reproductive effects, reduction in non-target organisms
Bromoxynil	C	R(L)	Mutagenicity
Cadmium	C (1989)	R[c]	Oncogenicity, mutagenicity, teratogenicity, fetotoxicity, acute and chronic effects on humans
Captafol	C (1990)	C	Oncogenicity, acute and chronic effects on wildlife
Captan	C (1990)	C	Oncogenicity
Chlordane	C (1989)	C/S[b]	Oncogenicity
Cyanazine	A	R	Teratogenicity
Daminozide (Alar)	A	C (for food uses)[b]	Oncogenicity
DBCP	C (1981)	C	Oncogenicity, mutagenicity, reproductive effects, groundwater contamination
DDE (TDE)	C (1987)	C	Carcinogenicity, bioaccumulation, hazard to wildlife and other chronic effects
DDT	C (1987)	R (1973)	Carcinogenicity, bioaccumulation, hazard to wildlife and other chronic effects
2,4-D	A	R(L)[c]	Carcinogenicity
Dicofol	A	C (for all products >0.1% DDT)[b]	Ecological effects
Dieldrin	C (1987)	C	Carcinogenicity, bioaccumulation, hazard to wildlife
Dimethoate	A	C (all dusts[b]) R (all other uses)	Oncogenicity, mutagenicity
EDB	R (1984)	C[b]	Oncogenicity, mutagenicity, reproductive effects
Endrin	C (1987)	C	Oncogenicity, teratogenicity, reproduction in endangered and non-target species
Heptachlor	C (1989)	C/S[b]	Oncogenicity
Lindane	C (1990)	R,R(L)[c]	Oncogenicity, teratogenicity, reproductive effects, other chronic effects, acute toxicity
Monocrotophos		A	C[b] Impact on birdlife
Oxyfluorfen	A	R(L)	Oncogenicity
Parathion	A	R(L)	Acute toxicity, toxic to aquatic organisms
Sodium monofluro-acetate 108	R	C	Reduction in non-target and endangered species

Table 3.2 Continued.

Agricultural chemical	*Victorian registration*	*US registration*	*Properties and criteria of concern*[a]
2,4,5-TCP	R (1986)	C	Oncogenicity, fetotoxicity
Trifluralin	A	C (if >0.5 ppm)[b] R (L) (N-nitrosammines)	Oncogenicity, mutagenicity

A, Available; R, Restricted; R(L), Restricted: special labelling required; C/S, Cancelled (suspended); and C, Cancelled

[a] Drawn from US EPA (1990)

[b] Product registered in Victoria in 1991, but not in USA (C or S) = 9

[c] Product registered in USA in 1990, but not in Victoria (R) = 3

Source: Scott, 1991: table 24.4

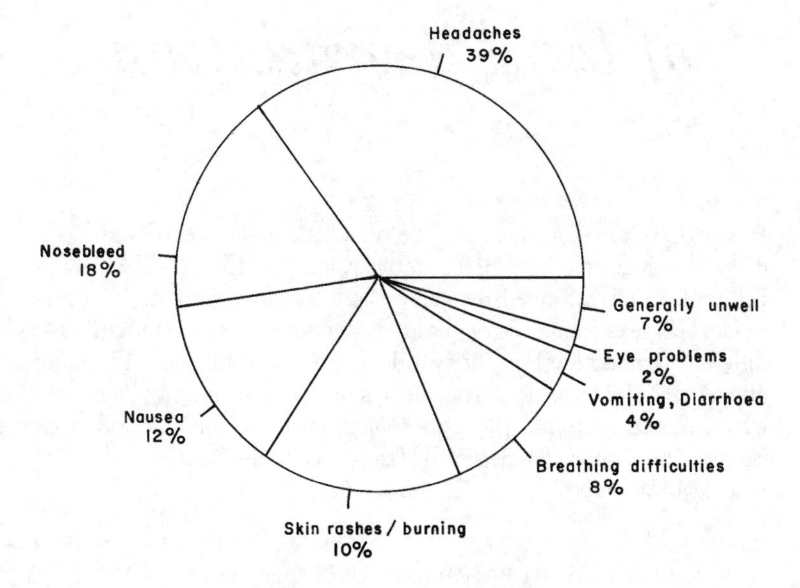

Figure 3.6 Symptoms of reported ill-effects from farm chemical use.
Source: Kondinin Group, 1993:28.

farmers choose ways that, in turn, cause further damage. That damage takes place not only on the land, or to the environment; but, as seen in the previous section, also to people—in this instance to their health.

But people are also seriously affected by land degradation through their economic activities and their social groupings; and this forms the subject matter of the next chapter.

■

4

Economic and social implications of land degradation

Perhaps the clearest and most direct economic consequences of land degradation are production losses, the costs of repairing the damage and, although perhaps less well recognised, the indirect costs associated with the use of pesticides. These economic costs in turn drive farmers to seek alternatives—different farming methods or farm locations, or a shift out of farming altogether. The latter in particular sets up a whole chain of cause and effect, impacting on the viability of country towns as economic and social centres. This chapter considers each of these broader implications of the rural land degradation problem in Australia.

Costs of repairing land degradation

The estimated 1975 costs of repairing the damage are presented in table 4.1. But these data present a puzzle. The total area requiring treatment with management practices and works adds up to 2 642 000 km2, which is far greater than the 815 000 km2 total for Australia in table 1.3. However, the mystery seems to be resolved by looking at Woods' (1983) table 7.3, which presents data on the areas of degradation requiring treatment in the arid zone (refer figure 1.2). That figure totals 1 850 000 km2 for Australia which, if added to the table 1.3 total, gets us very close to the total in table 4.1. Thus it appears that table 1.3 refers to the non-arid zone only. The lesson is that data need to be evaluated very carefully, and this point will be returned to below.

Table 4.1 Summary of treatment measures required and construction costs of works at June 1975.

Land-use category	*Australia*	*NSW*	*Vic.*	*Qld*	*SA*	*WA*	*Tas.*	*NT*	*ACT*
Area in use ('000 km^2)	5160	638	168	1620	571	1328	26	806	1.1
Area not requiring treatment ('000 km^2)	2493	41	69	929	175	746	25	507	0.30
Area requiring treatment with management practices only ('000 km^2)	1159	183	39	493	58	318	0.50	68	0.34
Area already treated with works[a] at June 1975 ('000 km^2)	26	16	n.a	4.2	1.4	4.3	–	0.015	0.17
Land still requiring treatment with works at June 1975									
area ('000 km^2)	1483	398	59	193	337	263	0.43	230	0.26
construction costs[b]									
LMW ($m)[c]	330	159	42	47	34	42	0.40	6.6	0.021
ECW ($m)[d]	345	172	80	63	16	6.4	1.0	5.6	0.75
Total ($m)	675	331	122	110	50	48	1.4	12	0.77
Land requiring[e] change of use at June 1975									
area ('000 km^2)	4.0	–	1.1	2.6	0.002	0.070	0.080	0.020	–
Annual maintenance (on required and existing works) ($m p.a.)	50	29	5.5	8.6	1.7	5.3	0.067	n.a	0.022

All values are approximate only and have therefore been rounded to two or three significant figures depending on the need for precision and the accuracy of the estimates.

a Works almost invariably require associated practices
b All costs at June 1975 prices
c LMW = Land Management Works
d ECW = Erosion Control Works
e Also included in costs and area of works (if these are necessary) under present land use. Reduction in land value resulting from change of land use has not been included.

Source: Woods, 1983: table 6.1.© Commonwealth of Australia, reproduced by permission.

Considering table 4.1 further, the total amount required to repair Australia's degraded lands in 1975 was estimated to be $675 million. Nearly half of this was required in New South Wales, followed (a lot further behind) by Victoria, Queensland, South Australia and Western Australia. These data are not in the same proportions as the area data, presumably reflecting the agricultural officers' estimates of the costs of treatment for different forms of land degradation, and also taking into account both the severity of that degradation and the varying degrees of difficulty in treating it.

The last row of table 4.1 provides estimates of the annual amount required to be spent on maintenance of required and existing works in 1975. These and the previously discussed estimates may be compared with the actual expenditure by state and territory soil conservation agencies in 1989/90 (table 4.2) and the Commonwealth contribution to soil conservation and Landcare (table 4.3). When inflation since 1975 is taken into account, it can be seen that all states are under-expending, by considerable amounts, the funds required to control land degradation.

How accurate are the data in table 4.1 twenty years later? The monetary values have changed considerably, due to inflation if for no other reason. Adjusting the 1975 estimated cost by the retail price index over

Table 4.2 Estimated expenditure by state/territory soil conservation agencies in 1989–90 ($ million).

	NSW	*Vic.*	*Qld*	*SA*	*WA*
Soil conservation organisation	54.0	10.999	7.649	3.460	9.329
Other state	–	–	–	1.400	0.880
Commonwealth (including NSCP and employment programs)	3.6	3.0	2.539	3.829	3.510
Other authorities	–	–	0.185	–	0.641
Total 1989–90	57.6	13.999	10.302	8.689	14.360
Total 1988–89	44.9	12.501	8.302	3.698	11.155

	Tas.	*NT*	*ACT*	*Total*
Soil conservation organisation	0.209	2.942	0.720	89.308
Other state	0.780	–	–	3.06
Commonwealth (including NSCP and employment programs)	0.606	0.455	0.352	17.891
Other authorities	–	–	–	0.826
Total 1989–90	1.595	3.397	1.072	111.014
Total 1988–89	1.219	3.489	1.565	86.829

Source: ABS, 1992: table 5.4.3. © Commonwealth of Australia, reproduced by permission.

Table 4.3 Commonwealth contribution to soil conservation and Landcare, 1974–75 to 1990–91.

Year	Budget [a] ($ million)	National	States	Admin	NSCP sub-programs			
					CLS[b] (dollars)	*Research*	*MPS[b]*	*PPET[b]*
1974–75[c]	0.154							
1975–76[c]	1.231							
1976–77[c]	0.180							
1977–78	0.095							
1978–79	0.105							
1979–80	–							
1980–81	–							
1981–82[d]	–							
1983–84[e]	1.000	400000	600000	nil				
1984–85	4.000	596549	3300000	103451				
1985–86	4.652	727735	3801500	137970				
1986–87	5.500	893190	4436000	163450				
1987–88	6.000	856727	4884000	165450				
1988–89[f]	10.634	1406224	5288256	386725	1558073	184348	605566	439921
1989–90[g]	23.184	2803609	16213843	355804	5436212	3353631	7721372	2506237
1990–91[g,h]	21.540	997452	18826716	136479	7559768	nil	9301015	2919164

Notes:

a Budget allocations may not equal expenditure in some years due to carry over funds.

b Abbreviations: CLS:NSCP Community Landcare Support Sub-program; MPS:NSCP Major Program Support Sub-program; PPET:NSCP Public Participation, Education and Training Sub-program.

c In addition, some $4.0 million were provided to state soil conservation agencies between 1966–67 and 1976–77 through the Commonwealth Extension Services Grants program.

d In 1981–82, a sum of $3.0 million per annum was added to General Purpose Grants to the states for soil conservation purposes.

e In 1983, the National Soil Conservation Program (NSCP) was established with two components — states and national.

f In 1988–89, the NSCP moved from two components, states and national, to a program-based approach involving four sub-programs. As there was a transitional phase in 1988–89 to the new sub-programs, projects were funded under the states and national components as well as the new sub-programs.

g The distribution of funds is presented as sub-program figures and also as national (other organisations) and states' figures.

h In 1990–91, the research sub-program of NSCP ($4.086) was transfered to the Land and Water Resources Research and Development Corporation (LWRRDC).

Source: ABS, 1992: table 5.4.4. © Commonwealth of Australia, reproduced by permission.

the intervening period yields a 1989 figure of $2.268 billion (Australian Bureau of Statistics, 1991:523). (1989 is the most recent date for which the long-term linked series of the Australian Retail Price Index is available: the more current Consumer Price Index (CPI) data do not extend back to 1975. However, the June 1989/90 'all groups CPI' of 100 had increased to 107.3 by June 1991/92, further inflating the 1975 repair cost figure to $2.430 billion by the latter date.) Several other estimates of the cost of repairing land degradation have been published, including those by Eckersley (1989), House of Representatives (1989) and Kirby and Blyth (1987). Their estimates range around $2 billion and are generally based on the 1975/77 study: that is, they are not rigorously revised estimates.

PRODUCTION LOSSES

The main economic (as distinct from environmental) effect of land degradation is reduced productivity. Despite the introduction of new cultivars and farm machinery, the increased use of synthetic pesticides and fertilisers, and the availability of extensive, government-operated support structures, crop yields per hectare have shown very little improvement in recent decades. Figure 4.1 shows that average Australian wheat yields, for example, have not exceeded 1.4 tonnes per hectare (although fluctuating considerably due to seasonal conditions), in contrast to rising trends exhibited by other OECD countries, especially in Western Europe where average yields have increased to over 6 t/ha. As shown in figure 4.1, there has also been little improvement in the 'broadacre' countries of North America. With specific reference to Australia, this failure to improve average yields cannot be attributed solely to the opening up of less productive, marginal lands. As has been shown by McWilliam (1981), for example, neither average commercial yields nor record yields (as measured by field crop competitions) showed any significant change from 1960 to 1980—despite the fact that about 60 per cent of plant breeding research funds were spent on wheat.

A specific indication of the effects of land degradation on yields comes from the results of a long-term New South Wales wheatbelt study, which showed that the artificial removal of the top 15 cm of soil resulted in yield reductions of up to 33 per cent (Hamilton, 1970). A second study, on the lighter, sandy soils of Western Australia showed that wind can remove the equivalent of 2.7 t/ha of the fertile dust fraction in a total loss of 27 t/ha of soil. A severely mismanaged soil lost 200 t/ha, including 30 t of dust. The removal of this fertile fraction from the top 8 mm of soil reduced subsequent wheat yields by 12–25 per cent.

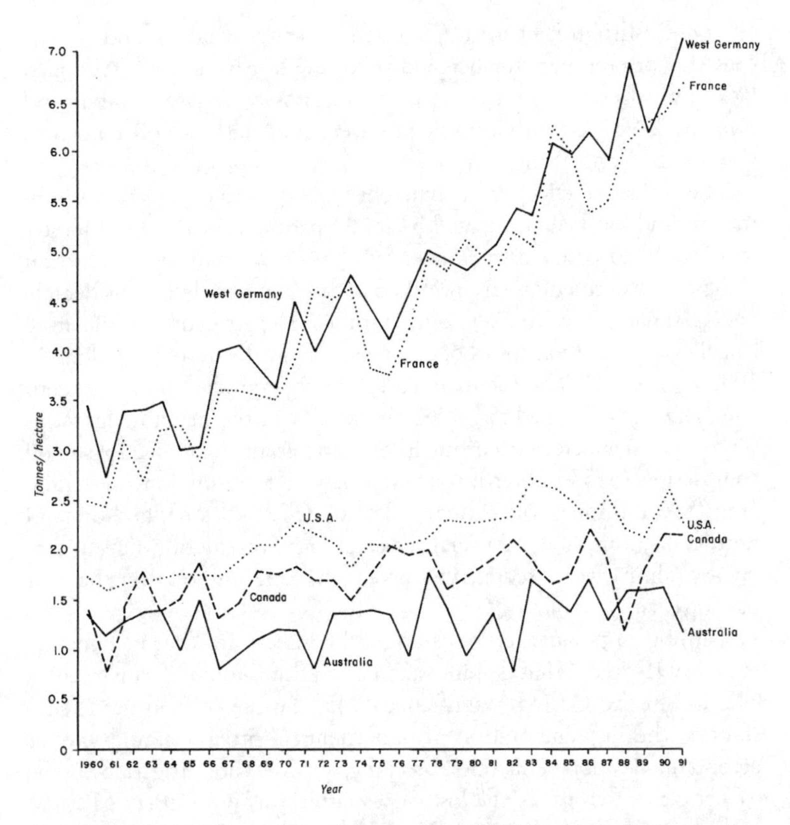

Figure 4.1 Wheat yields per hectare since 1960 for Australia and selected OECD countries.

Source of data: FAO Production Yearbooks, Rome.

Soils in this Mediterranean-type climate are particularly susceptible to erosion during the hot, dry summers; and it was shown that if summer grazing or cropping persisted for up to seven years in a row, productivity could be halved (Marsh and Carter, 1983).

For Australia as a whole, soil acidification has been estimated to result in production losses of $134 million per year (ABS, 1992), and the Western Australian *State of the Environment Report* considered that acidity had the potential to reduce production in more than 55 per cent of the agricultural areas of the state (Grant, 1992). Water repellence in the same state was estimated to cost $150 million in lost production in 1989 (Grant, 1992:93).

It needs to be stressed, however, that declines in (or static) productivity are generally due to a range of factors rather than any one cause. For example, Ellington et al. (1981) demonstrated that several factors caused poor crop performance and yellowing of wheat crops. Although lack of nitrogen is the usual agronomic diagnosis (see also Hamblin and Kyneur, 1993), contributing factors were root defects, soil hardpans, poor rotation and acidic soils.

The collective effect of the different types of land degradation is substantial and has been estimated by the Department of Primary Industry and Energy to cost Australia over $600 m per year in lost production (House of Representatives, 1989). But this is probably a considerable underestimate, since the Western Australian Department of Agriculture calculated production losses of a similar amount for that state alone in 1988 (figure 1.3). The total annual loss of gross production in Western Australia was estimated to be $609 million, with the major dollar losses (but not area affected) occurring in the agricultural areas, due to: subsoil compaction ($153 million); water repellence ($150 million); waterlogging ($90 million); soil structure decline ($70 million), and dryland salinity ($62 million). Annual losses of gross production due to secondary salinity were predicted to increase to $952 million over the next 30–50 years.

Turning to production losses caused by insects and weeds, Pimentel et al. (1991) found that despite a more than thirty-fold increase in pesticide use in the United States since 1945, the use of more toxic substances, the implementation of non-chemical programs, and various agronomic developments (crop breeding, fertilisers and irrigation), some 37 per cent of crops are *still* lost to pest attack (13% to insects, 12% to pathogens and 12% to weeds). The most significant increases were crop losses to insects, which almost doubled. In contrast, losses to weeds fell noticeably, while losses to pathogens increased only slightly.

There have been various attempts to assess the magnitude of these kinds of losses to pests in Australia, usually countered by suggestions that pesticide use *prevents* annual production losses of between $2 and $4 million, with a benefit-cost ratio of 4:1 (Senate Select Committee, 1990). This ratio is similar to that which has been proposed for the United States (Pimentel et al., 1980). However, if environmental and social costs are taken into account, the 'benefit' is substantially reduced (see below).

There are several reasons for these crop losses, many relating to a range of environmental pressures and disregard of ecological principles. Some reasons include: loss of natural enemies; increased resistance of pests to pesticides; use of more susceptible crop varieties (made more attractive to pests through breeding or the use of fertilisers); use of monocultures, which increase vulnerability to pest attack; retention of crop residues, favouring certain pests; reduced consumer tolerance for cosmetic damage to food, and neglect of good management practices (such as crop rotations) (Pimentel and Edwards, 1982).

COSTS OF PESTICIDES

In 1991, Australian farmers spent over $600 million on pesticides (table 4.4). These chemicals were used to treat about 5000 economically deleterious species of pests, weeds and diseases that affect over 200 host crops (Senate Select Committee, 1990). Herbicides account for nearly 70 per cent of all pesticides used (ABARE, 1992:209) and are applied mainly to broadscale crops, especially in association with minimum tillage practices. In contrast, the horticultural industries, sugar cane and especially cotton, are the main users of fungicides and insecticides. In addition, nearly $300 million are spent annually on veterinary chemicals (table 4.4).

Average farm costs for pesticides and veterinary chemicals generally account for between 3 per cent and 10 per cent of total farm expenditure, varying according to farm type (table 4.4). In some farm operations, such as broadacre cereals, annual expenditure on herbicides alone represents one of the major farm inputs, ranging from $10000 to $20000 per farm.

Indirect costs

Investment in pesticide controls has been shown to provide substantial economic benefits, with a return of $3 to $5 for every $1 invested in the United States. However, these figures do not reflect the complex costs, which include human poisonings, reduction of wildlife populations, fish and livestock losses, destruction of susceptible crops and natural vegetation, bee colony losses, evolved resistances to pesticides, creation of secondary pest problems, and the various administrative, insurance and monitoring costs (Pimentel et al., 1980). More recently, Pimentel et al. (1991) have indicated that pest control measures now cost the United States approximately $4.1 billion every year; and this figure does not include indirect environmental and public health costs. They conservatively total the latter at more than $2.2 billion per year, suggesting that the actual cost could be double that amount. If this is so, then the above benefit-cost ratio is substantially reduced.

AGRICULTURAL EXPANSION

A consequence of production losses, whether caused by land degradation, pest attack or other factors, is that farmers attempt to increase

Table 4.4 Average per farm business expenditure on farm chemicals by industry 1991–92; also indicating these as proportions of total farm costs, and with some Canadian comparisons.

<table>
<tr><th>Farm type</th><th>Crop and pasture chemicals ($'000)</em></th><th><em>Veterinary chemicals ($'000)</th><th>Total chemical[a] expenditure as % of total cash costs (1988–89)</th><th>Pesticides[b] Canada—av. cost per farm (1986) ($'000)</th></tr>
<tr><td>Poultry</td><td>2.0</td><td>3.8</td><td></td><td>3.4</td></tr>
<tr><td>Fruit</td><td>6.0</td><td>0.1</td><td rowspan="2">6.8</td><td>3.1</td></tr>
<tr><td>Vegetable</td><td>12.2</td><td>0.9</td><td>4.9</td></tr>
<tr><td>Cereal/oil</td><td>14.6</td><td>0.9</td><td>10.3</td><td>4.1 (wheat)</td></tr>
<tr><td>Sheep/cereal</td><td>10.7</td><td>2.1</td><td rowspan="2">8.1</td><td>5.3 (small grain)</td></tr>
<tr><td>Meat/cereal</td><td>5.6</td><td>2.3</td><td></td></tr>
<tr><td>Sheep/cattle</td><td>1.6</td><td>5.3</td><td>6.9</td><td>4.3 (livestock combination)</td></tr>
<tr><td>Sheep</td><td>1.0</td><td>3.0</td><td>6.8</td><td></td></tr>
<tr><td>Meat cattle</td><td>0.7</td><td>2.9</td><td>5.1</td><td>1.9</td></tr>
<tr><td>Milk cattle</td><td>0.6</td><td>5.0</td><td>2.9</td><td>1.6</td></tr>
<tr><td>Pigs</td><td>4.6</td><td>7.7</td><td></td><td>3.7</td></tr>
<tr><td>Sugar</td><td>2.5</td><td>0.1</td><td></td><td></td></tr>
<tr><td>Cotton</td><td>140.5</td><td>3.4</td><td></td><td></td></tr>
<tr><td>Other</td><td>3.3</td><td>1.1</td><td></td><td></td></tr>
<tr><td>Aust. total average per farm</td><td>5.5</td><td>2.7</td><td>3.8[b] (Canada)</td><td></td></tr>
<tr><td>Aust. total farm expenditure</td><td>$604</td><td>$298</td><td></td><td></td></tr>
<tr><td>[b] Canada total farm expenditure</td><td>$726</td><td></td><td></td><td></td></tr>
</table>

Sources: Australian Bureau of Statistics, Agricultural Industries Financial Statistics, Cat. No. 7567.0; a Bureau of Rural Resources, 1989: table 3, p. 2869; b Government of Canada, 1991: from table 9.5, section 9.22.

total production. This can be done in one of two ways. The first is by expanding the area under agriculture.

Farm sizes have increased throughout Australia's farming areas. In one decade alone, the average size of wheat farms increased by approximately 30 per cent from 1144 ha in 1968/69 to 1519 ha in 1978/79 (Lewis, 1981), albeit with some very marked regional and sectoral (type of farming) variations.

However, there is very little new land available for conversion to agricultural purposes, although Nix (1988) believes the country could support at least double its present population through alterations to trade and dietary patterns, land use and technology. In 1990, Western Australia put a stop to any further expansion of agriculture into even more marginal areas of the southwestern wheatbelt (McDonald,

1991:235), although clearing within the existing agricultural areas continues. Northern Australia and parts of Queensland are the main areas where new land is still available. But because of the duration of the dry period for some eight months or more each year, the isolation, and the uncomfortable climate for many people, especially during summer, the expansion of agriculture in the north will be slow.

Thus, given the general unavailability around Australia of new land for farming, farmers wishing to increase the size of their holdings have had to purchase existing farms. As the cost/price squeeze (chapter 6) intensified in recent years, farmers were actively encouraged by banks, agricultural agencies and advisers to 'get big or get out'. However, by the mid- to late 1980s, with interest rates increasing to 23 per cent or more, many farmers found themselves in financial strife. Expansion of farm size—the need for which was at least partly triggered by the failure to improve yields as a result of land degradation—has therefore become a less viable option for many farmers.

The second approach to combating production losses is by increasing farm productivity—output per unit area. One means of doing this is by irrigating.

Some consequences of irrigation farming

Various Murray/Darling Basin studies have amply illustrated the adverse effects of irrigation on the quality and productivity of agricultural land. Irrigation farming in northern areas has not been particularly successful either, due to ravages by birds and insects, and questionable economics and politics, despite concerted attempts over a number of years to find appropriate crops. The Camballin (WA), Tipperary and Humpty Doo (NT) projects were spectacular and expensive failures.

The Ord irrigation scheme in Western Australia still survives. First established in 1962 (before construction of the main dam in the late 1960s–early 1970s), and based on a potential irrigated area of 70 000 ha and assumptions of high crop yields, the scheme has experimented with a wide range of crops over the years. Initial plans to grow sugar were abandoned when world prices fell sharply. Instead, the first crop was cotton, but this was abandoned in the early 1970s due to serious pest resistances to pesticides (mainly DDT) and extremely high production costs. A sorghum crop exported to Japan in the late 1970s cost the Western Australian government $150 000 in direct subsidies. In 1978 a government investigation concluded that

> 40 years of research and 17 years of farming experience have simply resulted in the construction of a large irrigation project in which the State invested nearly $100 million (about $12 million of which was recouped in charges) and on-

which it is impossible for farmers to make a satisfactory profit. (Joint Committee, 1979:19)

Geographer Joe Powell (1988:22) was more blunt, describing the project as 'a triumph of wishful thinking and outright chicanery, a financial blunder of monumental proportions . . . a gigantic and basically useless development.'

Economists have also strongly criticised this and similar irrigation projects for their economic weaknesses, noting that right from the start costs would be high on account of the projects' locations, distances from markets and ports, and high infrastructure and labour costs (Davidson, 1967; Davidson and Graham-Taylor, 1982). The Ord scheme could succeed only if yields were high, through economies of scale or by heavy subsidies. The latter course was adopted, probably more by default than planning.

Several environmental problems also caused concern. Heavy pesticide use on the Ord's crops resulted in environmental contamination, and in 1972 high DDT residues in the meat resulted in an export ban being placed on beef which had been grazed on irrigated pastures. Wildlife in the area was also subjected to unacceptably high levels of pesticide residues, and the disappearance of one species of shrimp in the Ord delta was probably due to this cause. High concentrations of pesticide residues were found in the estuarine muds. Weed growth in irrigation channels and algal blooms in Lake Argyle have been reported. Little attention was given to the potential for siltation, which almost certainly has occurred given the run-down nature of many of the pastoral catchment areas (some of which have undergone rehabilitation). In 1988 the *The West Australian* newspaper reported the State Water Authority's district engineer as stating that 24 million tonnes of silt were filling the reservoir each year (18 June 1988:35). At this rate the dam's capacity would be reduced by 40 per cent within 100 years.

Any intensification of development in the area (at present confined to specialised grain, legume and fruit and vegetable crops) will undoubtedly exacerbate these problems. If the records of the Murray/Darling and overseas irrigation schemes are anything to go by, then other problems such as waterlogging, soil structural decline, salinisation, siltation, eutrophication and disease problems are entirely predictable (Brown, 1981; Tolba et al., 1992).

Nonetheless, in April 1993 the new Liberal government in Western Australia was reported as stating that 'the scheme had finally turned the corner' and that the prospects for a sugar industry were good. This is despite the fact that after 30 years only 14000 ha of the expected 70000 ha had been developed, with accumulated losses of $400 million (*The West Australian*, 12 April 1993).

Increasing farming intensity

Other means of improving farm productivity include adopting conservation farming practices in order to improve soil fertility (see chapter 7), and increasing the intensity of farming. Means of achieving the latter include the use of higher inputs of fertiliser, continuous cropping, high-yield varieties, irrigation (discussed above), and converting to corporate ownership to increase the availability of financial resources. The clear danger is that the intensification of farming may cancel out the beneficial effects of the adoption of conservation farming methods.

One particular response to the pressures exerted on farmers by the cost/price squeeze and land degradation was the shift to minimum tillage methods in broadacre cropping areas: the environmental implications of this were discussed in chapter 3.

OFF-FARM LAND DEGRADATION

While no doubt self-evident, it is probably worth emphasising that the damage caused by land degradation does not stop at the farm gate. Increased flooding and sedimentation affect streams, rivers, dams, fences, highways, railways, bridges, neighbouring properties, towns and cities. A minor example is presented by the Jerramungup Shire of Western Australia, which spent more than $40 000 in 1983 to remove sand blown on to roads from wind-eroded farms (Carder and Humphry, 1983). Those funds could have been used for more constructive purposes. On top of that, the on-farm losses to wind erosion were costed at approximately $18 000 (Marsh and Carter, 1983).

Indeed, it has been proposed generally that the off-site effects of farm erosion could be as much as a half to one order of magnitude greater than the on-farm effects of soil productivity losses (OECD, 1989).

Water quality

The loss of water quality, and the expense of restoring that quality, are other off-farm costs of land degradation. Water quality may be diminished by the effects of erosion (sedimentation and turbidity), nutrient enrichment (from fertilisers, manures and sewage), nitrate pollution, solid wastes (organic matter and litter) and loss of amenity (excellent examples are provided in WAWRC, 1992). Some of the adverse effects of eutrophication and nitrates, for example, have been discussed in chapter 3. It is important to recognise that off-farm effects may occur at some distance from the source and be of considerably greater significance than the Jerramungup example cited above. For example, sedimentation from Queensland cattle lands, comprising an estimated 15 million tonnes per year, is believed to be affecting parts of the Great Barrier Reef, impacting on fish breeding and causing algal blooms. Nutrients have been swept as far as 200 km offshore (CSIRO, 1990; *The Australian*, 13 April 1993).

According to the Department of Arts, Heritage and Environment (1986b), water quality problems in Australia include:

- *salinity* in many coastal-flowing rivers in Queensland, northern New South Wales, the Murray/Murrumbidgee system, western Victoria, southeastern South Australia, southwestern Western Australia, and in ground water in most of the country;
- *turbidity* in surface waters in most of New South Wales and Victoria, and a small area south of Perth;
- *undesirable colour* in some Queensland rivers, the Murray and many of its tributaries, and streams in western Tasmania;
- *heavy metals* in streams at a number of sites, mostly related to mining operations on the eastern seaboard, and in Tasmania, and
- *nutrients* in south eastern Queensland, much of New South Wales and Victoria, and in south western Western Australia.

Reduced water flow, acidification and other chemical pollutants, including pesticide residues, could be added to the list.

Table 4.5, from Garman (1983) in Department of Arts, Heritage and Environment (1986b), lists problems likely to affect water resources to the year 2000. Those increasing in surface waters are microbiological problems, excess nutrients leading to algal blooms and eutrophication, salinity and turbidity. Three of those problems are of national significance (salinity being regional in its extent), and all are considered to have a high degree of severity. The following section elaborates on salinity: this and eutrophication (discussed in chapter 3) are probably the two most serious water quality problems in Australia.

Water salinity

Increasing water salinity is a problem for a range of reasons, reflecting the wide range of water users. The reasons include ecological effects, yet there is a paucity of data on the effects of increasing stream (and dam or lake) salinity on indigenous, aquatic flora and fauna. This appears to reflect the attitudes that many Australians have to the biotic environment—that it is a resource to be used, as distinct from a complex, interconnected array of organisms which have an inherent right to existence (the latter viewpoint being that espoused by 'deep ecologists').

The significance of increased water salinity for various water uses is indicated in table 4.6. The data are indicative only, and are presented

Table 4.5 Problems likely to affect water resources to the year 2000.

	Extent	*Severity*	*Trend*
SURFACE WATER			
aquatic weeds	regional	moderate	steady
colour	regional	moderate	steady
corrosion	local	moderate	steady
hardness	regional	moderate	steady
heavy metals	local	low	decreasing
iron and manganese	local	moderate	steady
microbiological	**national**	**high**	**increasing**
nutrients—algae	**national**	**high**	**increasing**
oils and greases	local	low	steady
organics (natural)	regional	low	steady
oxygen demanding substances	regional	low	steady
pesticides	local	low	decreasing
pH	regional	low	steady
salinity	**regional**	**high**	**increasing**
turbidity	**national**	**high**	**increasing**
temperature	local	moderate	steady
taste	local	moderate	increasing
GROUNDWATER			
corrosion	local	moderate	steady
fluoride	regional	low	steady
hardness	national	moderate	steady
nitrate	regional	low	increasing
pH	local	low	steady
salinity	national	moderate	increasing
temperature	regional	low	steady
toxic pollutants	local	moderate	steady
taste	local	low	steady
radioactivity	regional	low	steady

Source: Department of Arts, Heritage and Environment, 1986b: table 2.6. © Commonwealth of Australia; reproduced by permission.

partly to point out the difficulty of identifying threshold values. For example, the tolerance of sheep to concentrations of soluble salts in their drinking water depends on a range of factors including their age and sex, whether a ewe is pregnant, and the nature of the solid feed available. A sheep can tolerate higher salt concentrations in drinking water if it is eating green feed than if it is subjected to a diet of dry grass or, worse, salt-tolerant shrubs.

Thresholds of salt concentrations in water used to irrigate crops also vary, this time depending on the method of irrigation (flood, overhead spray or trickle) and especially on the drainage efficiency of the irrigated soil. Well-drained soils in irrigation agriculture permit water with relatively high salt concentrations to be used, as the water can be flushed below the rooting depth of the crops, thereby preventing the build up of the salts in the root zone. As table 4.6 shows, different plants have different levels of tolerance to salt; and that tolerance level will, in turn, also depend on the age of the crop. For example, most plants are much more susceptible to salt at the germination stage than after they have become established.

The major area in Australia affected by increasing surface water salinity is the Murray/Murrumbidgee system (figure 4.2). Poor irrigation practices in particular, notably inadequate drainage, have resulted in salt concentrations increasing in groundwaters and soils, with groundwaters approaching the surface and in turn adversely affecting the irrigated

Table 4.6 Examples of tolerance to salt concentrations in water for selected plants, animals and human activities.

Total soluble salts concentration in mg L⁻¹	*Use of water*
200	pulp and paper manufacture
250	high grade steel production, automotive and electrical industries
450	clover[a]
500	WHO standard for human consumption
600	stone fruits, citrus[a]
1050	vines[a]
1500	WHO maximum permissible limit for human consumption
2000	beans[b]
2500	carrots, onions, clovers[b]
3000	lettuce[b], poultry
3800	potato, sweet potato, corn, flax, broadbean[b]
4400	pigs
5000	alfalfa, broccoli, tomato, spinach, rice[b]
5800	soybean[b]
7000	horses
7300	dairy cattle
7500	beets, sorghum[b]
9000	safflower, wheat[b]
10200	sugarbeets, cotton[b], beef cattle, lambs, weaners, ewes in milk
11200	barley[b]
11600	bermuda grass, tall wheat grass, crested wheat grass
13200	adult sheep
18000	adult sheep on green grass
36000	mean salinity of seawater in Cockburn Sound, WA
55000	camels

a The water quality criteria for clover, stone fruits, citrus and vines refer to the quality of irrigation water used and should not be confused with the soil-water quality criteria described in note [b]. The latter criterion is in many ways more meaningful since the effects of irrigation can vary widely depending on the method of application, the duration of irrigation and the quality of soil drainage.

b These are approximate values, from saturated soil extracts during the period of rapid plant growth and maturation, at which 50% yield reductions can be expected.

Source: From Conacher, 1982: table 9.2

crops (Murray/Darling Basin Ministerial Council, 1989). The Murray River has essentially become a drain, transporting increasingly salty water away from the irrigation areas, but with each downstream area progressively having to use increasingly salty water (figure 4.3). In terms of numbers of people and activities, the place most adversely affected by this process is the city of Adelaide, which still relies heavily on the Murray for its water supply. Trucks loaded with blue plastic containers delivering fresh water to offices are a common sight in the city, and many households have three taps—the third yielding potable rainwater from a tank filled by runoff from the roof.

The second region severely affected by increased water (as well as soil) salinity is the southwest of Western Australia. Salt concentrations in all the major streams in this region have increased since European settlement, and their waters are now unfit for human consumption. This is because the headwaters of the major streams and rivers are located beyond the forested areas, in the cleared, agricultural areas. Indeed, it has been said that nowhere else on earth has an area of this size (about 22 million

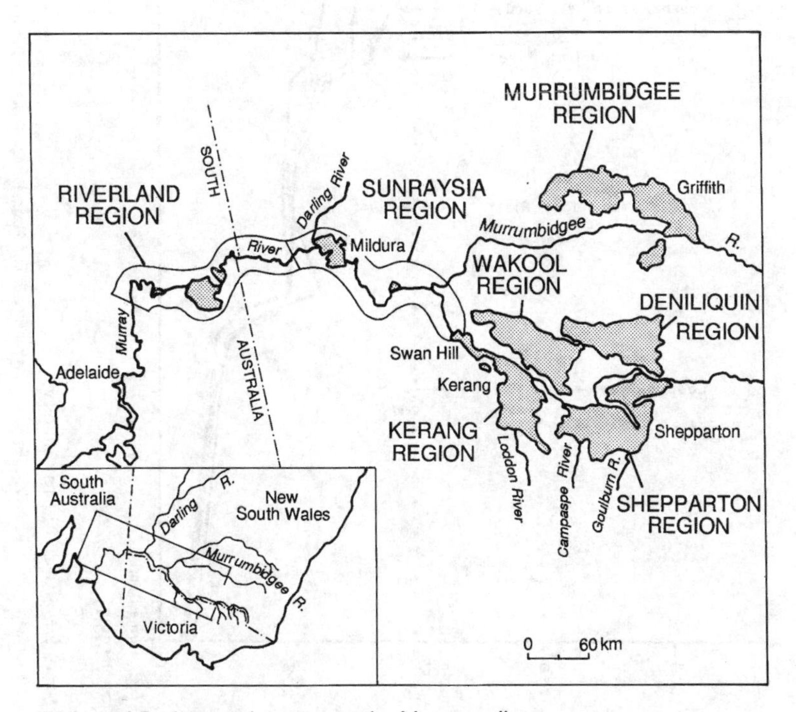

Figure 4.2 Irrigated regions in the Murray valley.

Source: Pigram, 1986: fig. 8.6.

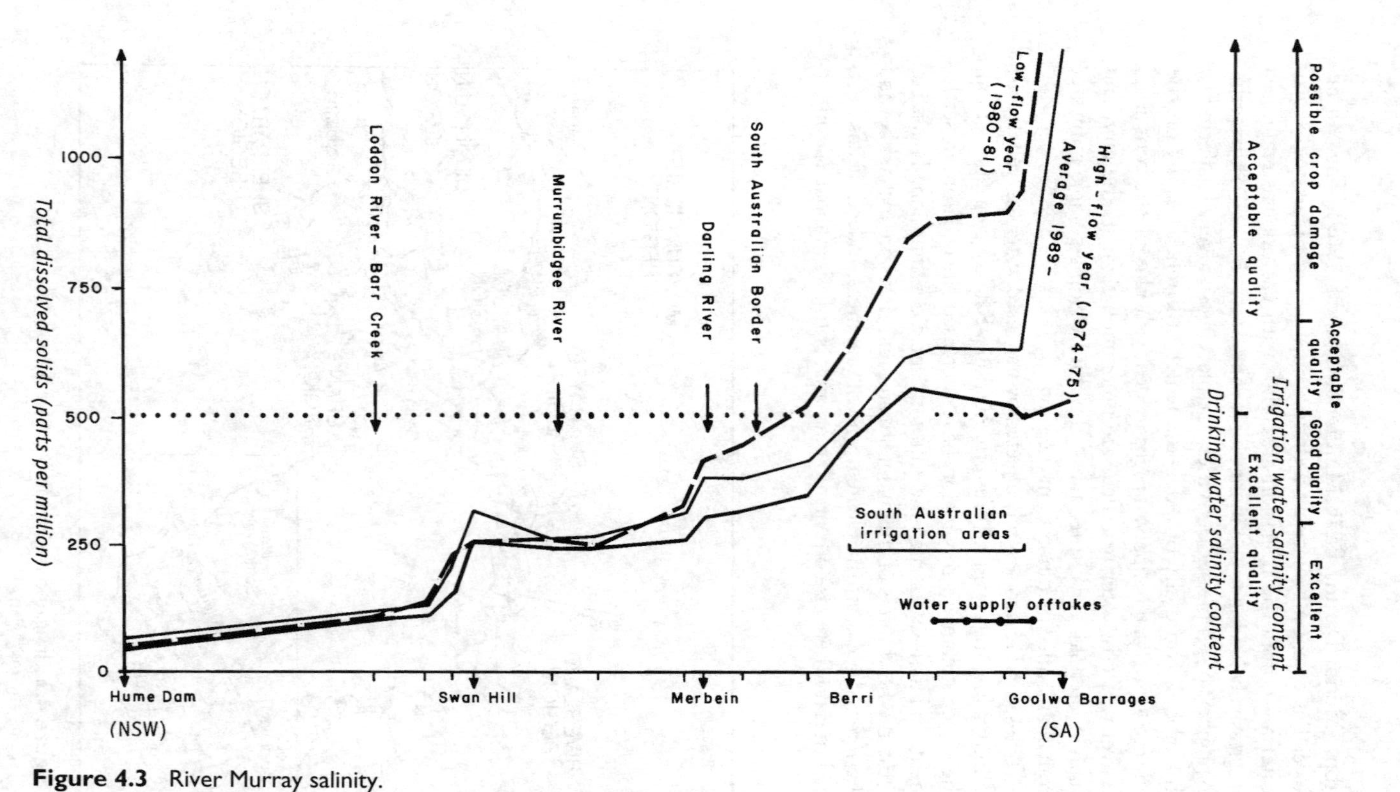

Figure 4.3 River Murray salinity.

Source: Modified from Green, 1983: fig. 5. © Commonwealth of Australia, reproduced by permission.

ha) been cleared of its native vegetation over such a short period of time (since 1829 but mostly in the twentieth century) *and* where nearly all surface waters have been rendered unfit for human consumption.

Unlike South Australia, however, there is no major population centre relying on these polluted streams and rivers for its water supply. Nevertheless, the Western Australian problem is exacerbated by the fact that the regionally extensive groundwaters are also naturally saline, although they are usually within the tolerance limits for stock.

LAND-USE CONFLICTS

As has been implied by the above discussions on water quality, which impacts on distant as well as local communities, and several of the adverse implications of the use of agricultural chemicals, the intensification and expansion of agricultural practices exacerbate land-use conflicts. As has been argued, this is at least partly the result of land degradation. Two broad groups of conflicts may be identified. One is characterised by competition for the same land, and the second involves land uses which conflict with one another at a distance.

Competition for agricultural land

There have been many examples of the first type of conflict—competition for land—since the earliest days of European settlement in Australia (Powell, 1976). Early declarations of state forests, national parks and various kinds of reserves (including Aboriginal reserves) were often a response to the extensive clearing of native vegetation for agricultural purposes.

This competition for land has been intensified more recently with the establishment of pine and eucalypt plantations on previously productive agricultural land. In 1990 there were more than 940000 ha of coniferous plantations in Australia, 230000 ha more than a decade earlier. In addition there were almost 100000 ha of broad-leaved plantations, mostly eucalypts, double the area only ten years previously (ABS, 1992: table 2.2.3). These figures compare with nearly 41 million ha of native forest, and another 63.5 million ha of woodland (with the latter shrinking more rapidly than the former) (ABS, 1992: table 2.3.1). While plantation forestry is to some extent intended to reduce pressures by timber and paper pulp interests on indigenous hardwood forests (especially the 'old growth' forests), it is ironic that other areas of native vegetation are still being cleared for agricultural expansion.

River salinisation is a second example, in that it has led to conflicts between agriculture and water catchment protection. For example, between 1976 and 1979 clearing bans were imposed by the state government on five river catchments in the southwest of Western Australia for this reason. In the Collie catchment, which supplies Wellington dam, farms were purchased by the (then) Public Works Department for reforestation in order to reduce salt concentrations in the dam. The total area affected by the clearing bans—that is, the catchments above the dams or dam sites—was a little over one million ha, of which 262 000 ha (of 398 000 ha in private ownership) had already been cleared (Public Works Department, 1979: tables 3 and 4). The bans caused considerable ill-feeling among the farmers and, along with other consequences, led to the acceleration in the rate of clearing of properties in other catchments which farmers feared might also come under similar bans. Even more ill-feeling was generated in 1993 when the Australian Taxation Office announced that (the often-inadequate) compensation payments to these farmers would be regarded as taxable income.

A third example concerns urban expansion and the proliferation of hobby farming. From the earliest days of settlement, urbanisation has conflicted with agriculture despite their close symbiotic relationship. Most towns and cities are located adjacent to waterways. As a consequence, areas of highly fertile, alluvial flats were among the first areas to be lost from agriculture. The search for rural 'retreats', which intensified from the 1970s onwards, has exacerbated the conflict.

In Victoria, it has been estimated that 28 per cent of 'prime' agricultural land suitable for intensive horticulture was lost to urbanisation, roads, industrialisation and supporting infrastructure between 1935 and 1983. Of the 75 000 ha remaining in 1983, a further 1900 ha were lost by 1989 (Scott, 1991:76–7).

Scott (1991) indicates that there are insufficient data to define precisely the extent of changes in land use away from agriculture generally across Victoria (this is also true for the rest of Australia). But the information which is available indicates that the greatest losses of agricultural land in the decade of the 1980s were experienced in municipalities close to Melbourne and a number of major regional centres. It was thought that these changes were most likely to result from property speculation, pressures for urban expansion, and the demand for urban/rural residential, rural retreat and small farm allotments. Such processes operate adjacent to all major cities.

Indirect land-use conflicts

Examples of the second group of land-use conflicts—land uses which conflict with one another at a distance—are also reasonably self-evident. Many of them concern the broader effects of land degradation on the wider community, as previously discussed (such as sedimentation, salinisation and eutrophication of waters at some distance from the farms which are responsible for the problems).

This category also includes the impact of other land uses on agriculture. In particular, urban expansion impacts on agriculture indirectly as well as directly: cities need water, building materials, transportation routes and energy as well as food. They also need to dispose of their waste. Thus areas adjacent to cities, as well as often being characterised by intensive market gardening, battery hen facilities, piggeries, orchards, stables and hobby farms, are also in demand for quarries, water supply dams, power stations, roads, railway marshalling yards and airports, and land for processing sewage and disposing solid and liquid wastes.

Air pollution and agriculture

Urban and industrial pollution of air, soil and water is also important in certain areas. Fluoride emissions—for example from alumina smelters and brickworks—cause necrosis of vegetation (a plant disease resulting from death of the tissue) and may affect stock health; nitrogen dioxide, mainly from vehicles, increases nitrogen levels in soils; ozone inhibits plant growth; acid precipitation lowers the pH of water bodies and soils; dioxins from burnt industrial wastes may affect organisms directly and indirectly; heavy metals from leaded petrol and industry contaminate soils and plants with potential effects on human health, and soot and dust add to atmospheric pollution (ABS, 1992: tables 3.3.1, 3.3.2 and 3.3.3; see also Ormrod, 1978; Tiller, 1989). In Japan, the effects of acid rain and other forms of air pollution are reducing wheat and rice crop yields by as much as 30 per cent (OECD, 1989), while in the United States an estimated 2–4 per cent of annual crop production is affected by air pollution, representing a loss of $1–2 billion (Skarby and Sellden, 1984).

Climate change

Predictions of climatic changes in Australia due to the enhanced 'greenhouse effect', and their effects on agriculture, are still uncertain: however, agriculture both contributes to and is affected by the greenhouse effect. First, agriculture contributes several greenhouse gases (CO_2, methane and nitrous oxides) which comprise between 10 and 30 per cent of greenhouse gas emissions. These gases come from clearing, burning and cultivation (CO_2), ruminant animals (methane) and fertilisers and legumes (nitrous oxides) (Pearman, 1988; Henderson–Sellers and Blong, 1989).

Second, agriculture is potentially affected by a possible global temperature rise and shifts of synoptic weather patterns, and increasing unpredictability in the behaviour of El Niño. The effects may be increased rainfall and storms in some regions (the northeast coast of Australia) and decreased rainfall and increased drought periods in others (southwestern Australia). Such changes could cause: the margins of wheat cropping to contract in some areas; the boundaries for tropical crops to extend further south; crops from deciduous trees to be adversely affected by a loss or reduction of the winter chill factor which is necessary to 'rest' those plants; erosion to increase under storm events or crops to experience greater stress under higher temperatures or drier conditions; the risk of fire to increase in drier areas, and a range of pests and diseases to enjoy more or less favourable temperature and moisture regimes, depending on location (Pittock, 1989; Parry, 1990; Wasson, 1990; ESD, 1991; Scott, 1991).

Ozone

In contrast to the uncertainty surrounding the 'greenhouse effect', the depletion of upper atmospheric ozone and the consequential increase in the receipt of ultra-violet radiation at the earth's surface is now a recognised fact. Yet the most recent official evaluation of Australia's environment (ABS, 1992) devotes fourteen pages (111–25) to the possible effects of 'greenhouse'-induced climatic changes—including their implications for agriculture—but nothing on the effects of increased ultraviolet radiation on crops and livestock. This is also true of Western Australia's *State of the Environment Report* (Grant, 1992). Indeed, that state's inquiry into land conservation in the agricultural region considered the implications of neither greenhouse climatic change nor ozone depletion (Select Committee into Land Conservation, 1990). The Victorian *1991 State of the Environment Report* (Scott, 1991) devotes an entire chapter to the relationships between agriculture and the greenhouse effect; but it, too, is silent on the topic of ozone depletion.

Tolba et al. (1992:34–5) note that because of the impact of UV–B radiation on some plants, and therefore on the functioning of ecosystems, ozone depletion in the upper atmosphere will reduce agricultural and fisheries productivity in the long term. They refer to UNEP (1989) in reporting that about half of plant species investigated were sensitive to enhanced UV–B radiation. Sensitive plants typically exhibit reduced growth and smaller leaves. In some cases, these plants also show changes in their chemical composition, which can affect food quality and the availability of mineral nutrients. For exposures simulating a 25 per cent depletion of the total ozone column, increased UV–B radiation was shown to reduce food yield of certain economically important varieties by up to 25 per cent.

Ironically, there is also a problem of *increased* ozone in the lower atmosphere, where it is a component of photochemical smog associated with automobile and industrial emissions. However, high ozone concentrations are not limited to urban areas. Long-range transport of precursors (nitrogen oxides and hydrocarbons which are reactive in sunlight) causes ozone episodes far from source regions. These precursors originate not only in urban and industrial areas but also in tropical regions where there is widespread burning of vegetation. The main photochemical oxidants, including ozone, are very reactive chemically and cause significant damage to vegetation (for example, Skarby and Sellden, 1984), materials and human health. They have been implicated as a factor in the widespread decline of European forests since 1970 (Tolba et al., 1992). In parts of southern England, ozone concentrations often exceed the 25 ppb recommended limit for Europe. At 50 ppb, ozone may reduce plant growth by 15 per cent (Emsley, 1992). Other examples of yield losses of some agricultural crops under experimental conditions are presented in table 4.7. However, experimental work has found that some plants respond to ozone with rapid genetic changes to cope with the pollutant (Emsley, 1992).

RURAL POPULATION DECLINE

A major change in rural Australia has been the marked loss of rural populations. This process has been going on for many years, initially in response to farm mechanisation and later to increased wage costs.

Table 4.7. Yield losses of some agricultural plant species exposed to O_3 under various experimental conditions.

Plant species	*O_3 concn. (g/m^3)*	*Exposure duration*	*Yield reduction % of control*
Alfalfa	200	7 h/d, 70 d	51, top dry wt
Alfalfa	200	2 h/d, 21 d	16, top dry wt
Pasture grass	180	4 h/d, 5 d/wk, 5wk	20, top dry wt
Ladino clover	200	6 h/d, 5 d	20, shoot dry wt
Soybean	200	6 h/d, 133 d	55, seed wt/plant
Sweet corn	200	6 h/d, 64 d	45, seed wt/plant
Sweet corn	400	3 h/d, 3 d/wk, 8wk	13, ear fresh wt
Wheat	400	4 h/d, 7 d	30, seed yield
Potato	400	3 h/d, every 2 wk, 120 d	25, tuber wt
Cotton	500	6 h/d, 2d/wk, 13wk	62, fibre dry wt

Source: Tingey (1984), unreferenced but cited in OECD, 1989: table 6.3

These early losses of farm labour are still reflected in many rural landscapes, where the old worker cottages now stand empty or, on some farms, restored for urban dwellers' farm holidays.

More recently, however, farm population declines can be attributed more to the decline in the number of farms (in parallel with the increase in farm sizes), largely through farm amalgamations (see also figures 6.4 and 6.5). The rural workforce declined by nearly 25 per cent in the thirty year period to 1985. Some 19 000 farmers left agriculture between 1970 and 1985 and continued to leave at the rate of 1–2 per cent each year during the 1980s (Lawrence, 1987; Ginnivan and Lees, 1991). Taylor (1991) reports that in Western Australia alone, 600 farms were lost in only two years in the mid-1980s.

There are many causes of this rural population decline. The rural–urban drift is a well-established demographic phenomenon worldwide, reflecting lifestyle and social choices and employment needs, among many factors. But in Australia, land degradation is one of those factors, albeit with a relatively complex chain connecting cause (land degradation) to effect (population decline).

Essentially, the argument states that land degradation results in static or declining yields. To increase total production, farmers must either increase the size of their properties (as many have) or increase the intensity of farming. The latter is a difficult alternative, given the limitations of Australia's environment. The argument is also complicated by the cost/price squeeze (chapter 6), where land degradation is both a *partial* cause (through reduced yields) and an effect (as farmers intensify their farming practices, with adverse effects on their soil and water resources). An outcome of the growing economic pressures on farmers has been increasing levels of farm indebtedness. Those who have been unable to trade their way out of debt have been forced off the land.

DECLINING COUNTRY TOWNS

As farm populations have decreased for the above and other reasons (which include reduced family sizes), so the populations of country town hinterlands have fallen. Kellehear (1989), for example, has estimated that about a third of Australia's country towns are in decline. The demand for goods and services provided by those centres with declining hinterlands decreased to the extent that in many cases businesses were no longer profitable and were forced to close. Lower farm incomes, high rural unemployment (as much as 30% in some areas) and ageing populations have compounded the problem. Farm machinery sales, for example, virtually halved between 1982 and 1986 (chapter 6). Banks closed smaller branches. Supermarkets in larger regional centres undermined smaller, family retailers in smaller towns.

Those developments have been exacerbated further by the attempts of federal and state governments to reduce their costs by reducing services—in particular, by closing many branch railway lines and centralising hospitals, schools and agricultural agencies in regional centres. To give an example: 30 000 railway workers lost their jobs in the 1980s decade, and another 45 000 have been predicted to be shed by 2001–02 (Taylor, 1991:260), with most of these losses occurring in Queensland (30%), New South Wales (30%) and Victoria (20%).

School closures have important implications for younger members of the community. As it is, only 7 per cent of boys and 10 per cent of girls in rural areas finishing year 12 go on to further studies, compared with 27 per cent in urban areas (Campbell, 1992). The loss of such crucial services drives younger families out of the affected communities.

These actions have led to further, 'multiplier' losses of employment and closures of businesses; many smaller country towns today are virtually ghost towns, while the main streets of larger centres are dotted with vacant shops. Other causal factors have been discussed in some detail by Bowie and Smailes (1988:251–2). They note that population declines took place mainly in farming areas and towns with populations of less than around 2500.

Social implications

The social effects of these changes have been profound for many rural people (Bryant, 1992), though they have been poorly studied or acted upon. Campbell (1992:38) makes the poignant observation that 'the rural sector is ageing, declining, stressed and going broke'. Economic deprivation and social pressures increase levels of violence, stress and ill health. Rural people suffer markedly higher levels of substance abuse, psychiatric disorders, stress-related and chronic illnesses than urban dwellers, and rural suicide rates increased dramatically over the 1980s (Cullen et al., 1989). The conditions are similar in rural areas of the United States (Dosman and Cockcroft, 1989).

These social impacts occur through a sense of increased isolation, deprivation and worthlessness (especially after several generations on the land); reduced personal interaction with neighbours; reduced schooling, social and employment opportunities for farm children; and vastly increased distances to be covered (and therefore time and cost) to obtain goods and services that urban dwellers take for granted.

Closely related to these social problems has been the underlying and growing rural debt problem (refer chapter 6). In 1985, the Bureau of Agricultural Economics estimated that 5 per cent of farmers (often the below-average managers or those on non-viable farms) were unable or unlikely to trade out of difficulties, and a further 20 per cent experienced financial difficulties (cited in Ginnivan and Lees, 1991). Bank foreclosures and loss of income (many farmers have had negative incomes for some years) take their toll on individual, family and community health and stability. A high percentage of farmers are increasingly seeking other sources of income off the farm to offset growing levels of poverty in preference to turning to welfare agencies for help.

Just as there are still many efficient, productive farms, some towns have stable and even growing populations through possessing employment-providing activities. In general, country towns around Australia that have escaped decline are those located on the coast or adjacent to metropolitan regions, or those associated with viable mining activities, retirement and tourism, more specialised forms of agriculture (horticulture and viticulture), or extensive rural subdivisions for hobby farms and alternative lifestylers (Bowie and Smailes, 1988; Burnley, 1988).

The previous chapters have defined the nature of rural land degradation and outlined its extent and growth globally (in chapter 1) but mainly in Australia (chapters 1 and 2). Broader implications of land degradation in relation to the use of agricultural chemicals (chapter 3) and its economic and social consequences (chapter 4) have been discussed.

The following two chapters deal with the direct and underlying causes of land degradation. Understanding the reasons for the problem is a prerequisite to doing something effective about it.

■

5

Direct causes of rural land degradation

The common threads underlying most of the land degradation problems discussed in the previous four chapters are agricultural and pastoral management practices. These practices interact with the processes operating in, and the properties of, the biophysical environment, resulting in detrimental changes to those processes and properties (some of which are discussed in McTainsh and Boughton, 1993).

Land management practices include the replacement of native vegetation with introduced or exotic crops and pastures, repeated cultivation, the use of heavy machinery, an increased reliance on synthetic chemicals, and the introduction of and overgrazing by introduced animals.

Biophysical processes are hydrological (and climatic), geomorphic, pedological and biological in nature. They include the interception and redirection of water movements by vegetation; translocation of soil materials by wind, overland flow, throughflow, groundwater, mass movement and leaching; loss of soil structure; the formation of subsoil hardpans; the development of soil toxicities; nutrient cycling, and the activities of soil biota.

The *properties* of the biophysical environment affected by the interaction of the land management practices on the biophysical processes concern the *quality* of Australia's soils, water and vegetation cover.

All the above are interrelated, making it very difficult to present a clear, sequential discussion of cause and effect. The approach taken here is to consider first the effects of clearing, fire and cultivation on ecological properties, ecosystems and soil properties. The effects of farming practices on biophysical processes and properties are then discussed, referring particularly to nutrient cycling, water movements and wind erosion.

CLEARING AND ECOLOGICAL PROPERTIES

The immediate and possibly most important effect of clearing on environmental properties is an ecological one, most particularly the loss of wildlife habitat (see further, Hobbs and Saunders, 1993; Paoletti and Pimentel, 1992). If clearing is extensive, the loss of habitat may result in species extinctions of both plants and animals, as described in chapter 2. Agriculture also contributes to species losses by draining wetlands, using narrowly based genetic materials, destabilising biochemical balances, using disruptive agricultural chemicals and introducing exotic plants and animals. Associated with the loss of species is the threat to gene pools, reducing the genetic resources for future foods, medicines and pesticides. For example, many arthropods (insects, spiders and mites) and micro-organisms (bacteria and fungi) are beneficial to agriculture in pest and disease control (as predators, parasites and pathogens), nutrient cycling (soil macro- and micro-fauna and flora), pollination (insects) and as sources of food for organisms higher in food chains. They are also important bio-indicators of changes in environmental conditions; and any species loss, reduction or dramatic increase in numbers generally signal some kind of environmental stress.

It is not necessary for forests and woodlands to be totally cleared to place plants and animals at risk. Forests are being subjected to increasingly intensive logging (both clear-felling and selective logging) and, together with woodlands and heathlands, deliberate burning. These activities threaten the viability of entire ecosystems, not only directly, but also indirectly, through the increased incidence of diseases. In fact, they become systems under stress.

Phytophthora cinnamomi, for example, is a microscopic soil-borne fungus which attacks plant roots, with the effect of reversing the osmotic pressure across the outer root cell membranes. The plant becomes unable to take up soil water and eventually dies. This disease kills over 400 species of plants in some 30 per cent of the native forests of southwestern Australia (Hussey and Wallace, 1993:70–1), and is also present in Gippsland, Victoria. The fungus moves naturally downslope in soil water; but its spread is accelerated considerably by wet conditions and the physical translocation of diseased soil on vehicle tyres or bulldozer tracks. Logging, the construction of unsealed roads, and the spreading of road gravel from quarries in infected areas, are the major causes of the rapid spread of the fungus, which is thought to have been introduced into Australia from Indonesia in the 1920s. Farmers may also have inadvertently spread the fungus during clearing, constructing and maintaining tracks and firebreaks, and moving stock. There are at least six other fungi (including canker fungus and a species of *Armillaria*) which kill a wide range of plant species in southwestern Australia's forests, woodlands and heathlands: susceptible plants include some exotics such as avocados, proteas and lupins as well as native species.

Weeds

The transition from useful plant to weed occurs as a result of the plant's introduction to an environment where its growth and reproduction rates are much greater than those of the native flora. That transition may not occur immediately. Changes in climatic patterns or land management practices may be required to trigger the expansion of some species. *Mimosa pigra*, for example, was introduced to Darwin in the late nineteenth century but did not begin to spread seriously until almost a hundred years later. Skeleton weed and prickly acacia (*Acacia nilotica*) are other examples. The acacia remained under control in the Mitchell grasslands of northern Queensland for forty years and then, in the mid-1970s, exploded in density following a change from sheep to cattle grazing, associated with some exceptionally wet years.

Weeds are spread in a number of ways (termed 'vectors'), which may be natural or otherwise (usually involving human interference). Natural vectors include wind (thistles, serrated tussock), water (dodder, Noogoora burr, mesquite, honey locust), native animals and birds (dodder in kangaroo dung, blackberry seeds by birds), rhizome spread (bracken fern), and through soil, floods, fire and drought (prickly pear, tiger pear and prickly acacia). Unnatural means of spreading weeds include: carrying seeds in the hair or on the skin, or in the dung, of domesticated or feral animals; the contamination of weed seeds in produce such as feed products, crops, hay and wool; on vehicles; through irrigation channels; soil movement through a wide range of disturbances—earthworks, power lines, roads and quarries; the dumping of rubbish into landfills or parks, or ballast or bilge from ships; deliberate planting of pasture species and ornamentals; escaped ornamentals or soil with other plants; in clothes or in soil on footwear, and through the aquarium and nursery trades.

Fire and ecosystem stress

Fire is another factor which places ecosystems under stress. It has long been a natural (through lightning strikes) and cultural (through use by Aborigines and early settlers) part of the Australian landscape (Hallam, 1975). Marked geomorphic and vegetative effects have resulted (Hughes and Sullivan, 1986), although Boon and Dodson (1992) have pointed out that fire is only one of several factors which induce vegetation shifts and other environmental responses.

Unlike past use, fire is now used deliberately as a management tool (particularly to reduce forest floor litter loads which may fuel 'holocaust' fires) at far greater frequencies and extent. Higher frequencies are well demonstrated by a study of *Xanthorrhoea* sp. in jarrah forest near Perth. During the one hundred and fifty years prior to European settlement, the plants had been burnt only one to three times—an average rate of once every 50 years. In the 150 years since European settlement, fires have occurred between 12 and 22 times, or once every 10 years (Lamont, cited in Hussey and Wallace, 1993:141). Such increased burning frequencies are responsible for some profound changes to natural ecosystems.

'Prescribed' or 'cool' fuel reduction burns are carried out in spring and autumn, usually by dropping incendiaries in carefully designed patterns from fixed-wing aircraft. The effects of such burns are numerous, and include adverse effects on nutrient cycling (discussed below) as well as some serious impacts on soil physical, chemical and biological properties and to forest ecosystems (Gill et al., 1981; Tomkins et al., 1991). The ABS (1992:44) states that 'the main threat to maintenance of species diversity in state forest areas (areas from which timber is harvested) is fuel reduction burning.' It is known, for example, that birds such as mallee fowl require habitats in which fire has been excluded for at least forty years. This is because they use the woodland litter to build the large nests in which their eggs are incubated. But prescribed burns are carried out with an average frequency of about once every seven years, ranging from perhaps every three years in very fire-sensitive areas, to once every nine years in areas of low risk.

Nutrient cycling

Figure 5.1 shows diagrammatically an undisturbed block of soil (a three-dimensional *pedon*) with its associated vegetation and root systems. Nutrients are stored in all parts of this system: in the leaves, bark, twigs, branches, stem and roots of the vegetation (including low shrubs and

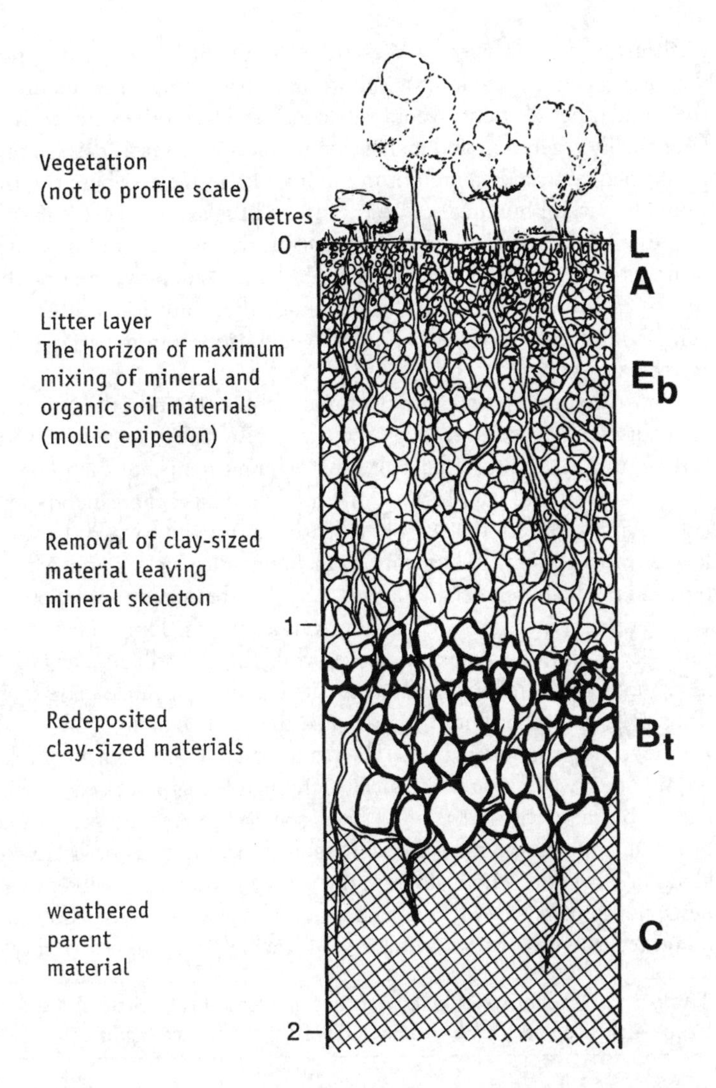

Figure 5.1 Diagrammatic illustration of an undisturbed pedon with its natural vegetation, indicating horizon development partly in response to vertical translocation of clay-sized soil materials.

ground covers as well as trees); in the litter layer on top of the mineral soil (incorporating fallen leaves, bark, twigs and branches in various stages of physical breakdown and decomposition by soil biota, fungi and bacteria), and in the various soil horizons.

Table 5.1 presents some data from 40-year-old karri forest on a red earth in the far southwest of Western Australia. It is stressed that these data are specific to that site and are indicative only; the amounts of these and other nutrients would vary considerably at other sites with different soils, vegetation and management practices. In fact, this is not an ideal example; the data are not from virgin karri forest (where the trees would be several hundreds of years old), and the site has been subjected to prescribed burning. Nutrient amounts in the biomass and litter layer of uncut forest would almost certainly be considerably greater than shown in table 5.1. On the other hand, there are few undisturbed (uncut, unburnt, ungrazed) natural vegetation communities left in Australia.

Under undisturbed conditions, these (table 5.1) and other nutrients are continually recycled through the system. Roots extract nutrients directly from the soil and from soil water. The nutrients are then pumped up (by suction and capillary action) into the plants, and incorporated (in varying proportions), by photosynthesis, into the wood, bark and leaves. Most Australian vegetation, including eucalypts, is not deciduous, and the plants shed bark, leaves, twigs and branches throughout the year (although there are very clear seasons for the shedding of bark).

Insects and their larvae chew leaves while they are still attached to the plant, and others (mainly macrofauna such as ants, termites, spiders, beetles and larvae) continue to do so once the plant remnants are on the ground, comprising forest litter. The plant materials are gradually broken down both physically and chemically (through being processed by saliva and in the gut of the insects), and the decomposition process is continued by soil microflora and microfauna, notably fungi and bacteria. Many soil biota, notably ants, physically mix the soil by constructing subterranean nests; nutrients released by decomposition processes are also dissolved in rainwater reaching the soil surface, and leached into the soil (Lobry de

Table 5.1 Amounts of selected nutrients in trees, shrubs, litter and soils on a red-earth site near Pemberton, southwestern Australia (kg/ha).

	N	*P*	*K*	*Ca*	*Mg*
Trees					
bole	108	10	147	550	102
whole	189	18	225	698	144
Shrubs	60	1.5	39	39	7
Total biomass	249	19.5	264	737	151
Litter	224	7	32	396	60
Soil					
total	7439	1718	6214	10224	5346
to 90 cm—extractable	–	842	471	3827	1027

Source: Hingston et al. 1979: table 5.3, p. 145.

Bruyn and Conacher, 1990). In these ways, nutrients are returned to the soil and then to the roots, where they are again taken up and used by the plant. The biotic population of a fertile soil is in fact phenomenal: according to Lee and Pankhurst (1992), soil biomass may exceed 20 t/ha, with a diversity which has been compared with that of coral reefs.

Obviously, vegetation clearance for whatever purpose completely disrupts the cycling of nutrients and in fact results in the removal of a significant proportion from the system altogether (figure 5.2). But it is not necessary to remove the vegetation to have this effect. Prescribed burning is *designed* to remove the litter layer. Forest and reserve managers perceive that layer not as an integral part of a nutrient cycle necessary for the health of the vegetation community, but as fuel. Preventing the accumulation of litter must result in the depletion of nutrients from the soil and eventually the plants, thereby affecting the health of the entire system.

The removal of plant and animal materials in the form of farm produce from agricultural systems is another important source of soil nutrient loss. Unless such losses are compensated for by the return of crop residues or the addition of animal manures, legumes or fertilisers, then the losses may become serious—particularly given the low inherent fertility of many Australian soils. Wild (1993), for example, has shown that nutrient removal in wheat grains (based on dry matter yield of 5 t/ha) amounts to 100, 20 and 28 kg/ha of nitrogen (N), phosphorus (P) and potassium (K) respectively, with additional losses occurring in the wheat straw. Under Australian conditions, 3 kg/ha of P is lost for every tonne of wheat grain produced (Scott, 1991).

WATER MOVEMENTS

In an undisturbed, vegetated environment, only a small proportion of rainwater reaches the soil surface directly. Most, if not all, is intercepted first by the canopy of the vegetation—whether that of trees, shrubs or ground covers—and then by the litter layer.

The amount of rain intercepted by the canopy varies considerably. It depends on the three-dimensional shape and size of the canopy, and the size and shape of the individual leaves. Small leaves, especially needle leaves of conifers, have a high surface area per canopy volume and are therefore able to retain much higher amounts of rain than, say, the large, fleshy leaves which characterise much of the vegetation in the wet tropics. However, the proportion of rain intercepted by canopies is dependent not only on such canopy characteristics, but also on the intensity

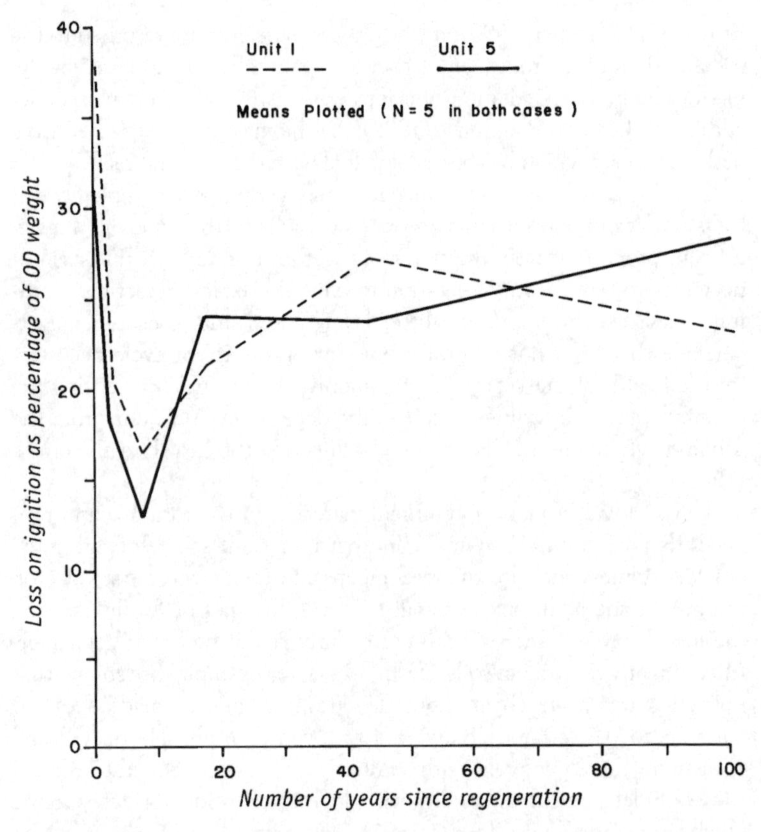

Data shown are loss of organic matter ('loss on ignition'), since organic matter content is generally highly correlated with the other major nutrients in these soils. Nutrient levels recovered as the forest regenerated, but did not regain the amounts measured under virgin forest (time zero). Unit I—interfluve sites; Unit 5—midslope positions.

Figure 5.2 **Loss of nutrients from surface (0–4 cm) soil materials following clear felling of the karri forests of southwestern Australia.**

Source: From Valentine, 1976: 16.

and duration of the rainfall event, and the amount of moisture already retained by the canopy prior to the rainfall event. In areas with cold seasons, the proportion of total precipitation which arrives in various frozen forms is also important. For these reasons, measurements on an annual rather than an individual storm basis provide a better indication of the role of vegetation in the water budget (table 5.2).

In other work reviewed by George (1991), interception of rainfall by *Eucalyptus* forests has been estimated to be between 10 and 25 per cent

Table 5.2 Some examples of median values of canopy interception as a percentage of annual or seasonal precipitation.

Vegetation	*% interception*
Deciduous forest	
all data	13
Coniferous forest	
rainfall only	22
rain and snow	28
European data only	35
North American data only	27
Taiwan	8
Alfalfa	
growing season	36
low vegetation development	22
Corn	
growing season	16
low vegetation development	3
Soybean	
growing season	15
low vegetation development	9
Oats	
growing season	7–23
low vegetation development	3
Spring wheat (growing season)	10–35
Rye (50–150 cm tall)	4–6

Various sources, from Dunne and Leopold 1978, tables 3-1, 3-2 and 3-3.

(Dunin and Mackay, 1982), 10 and 27 per cent (Raison and Khanna, 1982) and 10 and 13 per cent (Williamson et al., 1987). Nulsen et al. (1986) found interception to be as high as 35 per cent in mallee woodlands in a low rainfall (370 mm/yr), Western Australian wheatbelt environment.

As a rainfall event continues, the canopy eventually reaches its saturation capacity. Any further rain then either displaces water already retained by the canopy, or is redirected. The displaced or redirected rain reaches the ground in two main ways—by throughfall (or canopy drip, as termed by some writers) and stemflow. Canopy drip can be a significant geomorphic agent. Light, drizzly rain comes in small drops with no significant impact velocity. But when shed by leaves the drops may be large, and provided the canopy is tall enough (about 9 m), the drops will reach terminal velocity. Such drops, on reaching bare ground, may produce impact craters and displace quite large soil particles over some distance.

Stemflow refers to water which flows down the branches and then the trunk of the tree or shrub. Such flow may be quite concentrated and cause significant, albeit localised erosion around the base of trees. Measurements were made of stemflow and overland flow by Puvaneswaran in Western Australia's jarrah forests (reported in Puvaneswaran and Conacher, 1983). She found that stemflow had a much higher discharge per unit of affected area (by at least two orders of magnitude) than overland flow (table 5.3).

Under natural vegetation, water reaching the ground, whether directly or by throughfall or stemflow, is then intercepted by the litter layer. This both breaks up the direct physical impact of rain drops on the soil surface, and also acts as a sponge (depending on the humus content), storing a proportion of the water. The base of tree trunks, where stemflow often washes litter away and causes some soil erosion, is the main exception. As with canopies, the proportion of water retained by litter depends both on the characteristics of the litter layer (thickness and consistency) and on the intensity and duration of the rainfall event.

If a rainfall event continues, the storage capacity of the litter will be exceeded and water will infiltrate into the soil. Undisturbed surface soil horizons under natural vegetation are thoroughly mixed by soil biota—in Australia, primarily by ants and termites—and also have an intimate mixing of mineral soil particles with organic matter, particularly humus. These soils absorb water readily. In the drier and hotter parts of the continent, however, the organic matter content of the soil's A horizon is low, often less than 1 per cent, in comparison with that of more temperate regions (up to 6% or more). Soil textures are characteristically sandy (comprised predominantly of sand-sized—0.02–2.0 mm diameter—particles), as a result of preferential sorting by the soil biota, translocation of materials across the soil surface by surface wash (a combination of drop impact and overland flow), or removal of finer fractions down the soil by leaching. Thus in the drier areas as well, under natural conditions water also infiltrates readily, although some soils have water repellent properties (McGhie and Posner, 1980) or are naturally hard-setting.

Table 5.3 Comparison of volumes of water per area generated by overland flow and stemflow from seven rainfall events on four forested slopes in south-western Australia.

Slope	*Overland flow per 500 m² (ml)*	*Stemflow per tree crown area (range 3.5-7.8 m²) (ml)*
1	6109	4674
2	4033	7344
3	9740	7706
4	13456	10167

Source: Puvaneswaran and Conacher, 1983: tables 2 and 3.

At varying depths from the surface, the clay content (particle sizes of less than 0.002 mm diameter) of many soils increases fairly abruptly: these texture contrast soils are termed *duplex* soils, after Northcote's (1960) classification. The porosity of the soil decreases accordingly, as does the ability of the infiltrating water to continue its downward path, under the influence of gravity. A lateral component may therefore be introduced: these lateral, subsurface soil-water movements are termed *throughflow* (first defined by Kirkby and Chorley, 1967), or *perched*, ephemeral (if associated with individual rainfall events), seasonal, or perennial (if occurring all year round) aquifers. Some water continues to infiltrate vertically, however, and eventually reaches the permanently saturated zone, or groundwater. Figure 5.3 illustrates the various water movements discussed here: there is a clear relationship between these movements and the morphological and textural properties of the soil pedon shown in figure 5.1.

These water movements are crucial to the processes of soil formation, or *pedogenesis*. Soil-water carries dissolved and suspended materials, including plant nutrients, from the atmosphere, decomposed vegetation and the soil itself. The materials may be redeposited within the soil pedon, thereby changing the nature of the soil, or leached from the soil system into groundwater and eventually into rivers and the sea. Alternating wetting and drying conditions within and at the base of the soil provide optimum conditions for chemical weathering of the soil's parent material (which may be bedrock or sediment); and soil-water is of course crucial for soil life, both flora and fauna. Many of Australia's indigenous plant species in drier areas have deep tap roots, enabling the plants to draw on groundwater even during the long, hot, summer months. Water is returned to the atmosphere, leaving behind its dissolved load, by the transpiration of plants and by evaporation from the surface of plant canopies (a combined process termed *evapotranspiration*), and by direct evaporation from the soil.

Effects of replacing native vegetation with introduced plants

Removal of the native vegetation has a considerable effect on the water movements described above. The effect is determined by the nature of the plants which replace the native vegetation, and by the management practices—especially the length of time for which the soil surface is left bare. Burning of plant residues and trampling by stock also affect water movements.

Replacement of native vegetation, especially woodland or forest, with introduced crops and pastures results in reduced interception, reduced

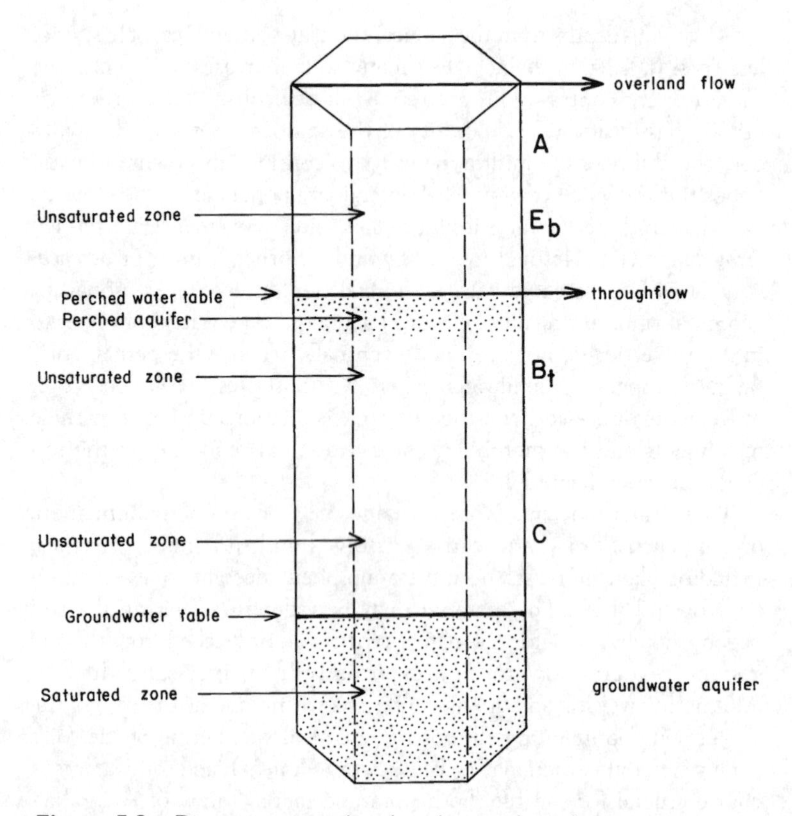

Figure 5.3 Diagrammatic soil pedon showing lateral water movements and aquifers, partly in response to the horizonation shown in figure 5.1.

throughfall, reduced stemflow, and reduced storage in canopies and litter. However, and as discussed by George (1991), the additional water which reaches the canopy of agricultural plants may also be intercepted and evaporated back into the atmosphere. Williamson et al. (1987) assume that this is zero, although Sedgley et al. (1981) quote an unsourced United States study reporting rates of up to 20 mm/yr. Rain interception by agricultural plants is nevertheless poor in comparison with *Eucalyptus* forests and woodlands, especially since the introduced annuals are present for only up to 6–8 months of the year.

Depending on management practices, more rain reaches the soil surface directly. Raindrops 'explode' on the soil surface, splashing particles into the air; on sloping surfaces, there is a net downslope translocation of these particles. Further, the finer particles tend to lodge in spaces between soil aggregates, or in faunal holes or between larger particles.

Raindrops also have a compacting effect on the surface. The combined effect is to 'seal' the soil surface, or at least to make it much less pervious to water.

Thus less water is intercepted, stored or infiltrated into the soil; and it follows that a much higher proportion of water than previously will accumulate on the soil surface, and fill hollows of the surface micro-topography. If the rainfall event continues long enough, these hollows fill and overflow, resulting in overland flow. Note that overland flow may also occur under undisturbed vegetation, but it requires much more intense rainfall events, or events of longer duration. While a rainfall event continues, the overland flow will be erosive due to the impact of raindrops, which create turbulence in the flowing water and also directly dislodge soil particles, which are then taken into suspension (and some in solution) by the overland flow and translocated downslope.

There is a vast literature on the mobilisation and translocation of soil materials by raindrop impact, overland flow and combinations of the two (often termed 'rainwash'), under both natural and human-modified conditions, such as the early review by Smith and Wischmeier (1962), the classic field experimental study by Emmett (1970), the more recent field- and laboratory-based work brought together in de Ploey (1983) and Bryan (1987), and chapters 2, 3 and 4 in Abrahams (1986).

ACCELERATED EROSION

The loss of soil materials following human disturbance is termed *accelerated* erosion, in that the rate of loss of materials has been accelerated as a result of human actions in changing the vegetation and possibly also disturbing the soil surface. Accelerated erosion is distinguished from natural erosion rates, or geomorphic erosion; although one wonders whether there is anywhere on the earth's surface which has not experienced *some* degree of modification due to human actions, however minor.

Accelerated erosion is frequently measured or predicted in terms of an American model known as the Universal Soil Loss Equation (USLE) (Wischmeier and Smith, 1978; see also Foster (1988:100), who refers to a 'typical' soil loss tolerance of 10 t/ha/yr). But, as pointed out by Rose (1988—who discusses other, more recent models) and earlier by Wischmeier himself (1976), empirical United States data used to 'solve' the equation should not be applied uncritically in other environments. Under Australian conditions, where natural soil formation rates appear to be extremely slow, the USLE may be inappropriate. Australian soils are often thin, light textured, of great age and inherently infertile, although there are exceptions in some of the more productive wet tropical and cool temperate regions.

Various studies of erosion rates, using erosion plots, have been undertaken around Australia under a variety of conditions. Broadly, soil loss rates increase from south to north and from inland to coastal locations. Losses under undisturbed conditions (such as forest) are negligible at 0–1 t/ha/yr; but can increase substantially after bushfires to 10–50 t/ha/yr. Soil loss rates are also very low under pasture (<1 t/ha/yr), increasing to 1–50 t/ha/yr under various crops, and ranging up to 100 t/ha/yr under bare fallow. Under some conditions, and especially where high intensity storms impact on bare soil surfaces, individual events have caused losses of several hundred tonnes per hectare (up to 700 t) in parts of Queensland, New South Wales and Western Australia (Edwards, 1993). At such high rates of soil loss, productive land can become useless within a few decades.

While significant soil losses have been suggested for pre-European settlement periods (Hughes and Sullivan, 1986), erosion rates undoubtedly accelerated after European settlement. Both findings are supported by the long-term data provided by Wasson and Galloway (1984), who presented data from the Broken Hill area in New South Wales (current annual rainfall about 200 mm). They obtained erosion rate data by measuring the amounts of sediments accumulated in an alluvial fan and, after 1915, in a reservoir.

They found that erosion rates were 'high' (83 mm/1000 yr) between 6000 and 3000 before the present (BP) but 'very low' (0.5–4.0 mm/1000 yr) from 3000 BP to 1850 AD. The high erosion phase was attributed to a 'warm and wet' climate during that period. Unfortunately, no estimate is given as to what they considered the rainfall to have been. The 'very low' erosion period is attributed to a drier climate than the present regime. After 1850, erosion rates increased dramatically, to a maximum of 290 mm/1000 yr between 1915 and 1941. Rates then dropped to around 109 mm/1000 yr, which is the present rate. The dramatic increase after 1850 was attributed to human impacts—through grazing of introduced stock, clearing, rabbits and mining. Improved catchment management was thought to be responsible for the decline in erosion rates after 1941.

Australian estimates on soil formation rates appear to vary widely from region to region (and author to author), from 2 to 200 mm/1000 years; however, Edwards (1991) notes that rates of up to 20 mm/1000 years are probably more appropriate. These data provide a useful comparison with the erosion rates cited above.

It should be noted that the beneficial effects of improved management practices identified by Wasson and Galloway make the important point that human actions may also reduce erosion rates. Too often, it is assumed that the effects of human actions are always adverse. For example, the clearing of trees removes the erosive (albeit highly localised) stemflow; and Puvaneswaran and Conacher's work (1983) in the jarrah forests of southwestern Australia demonstrated that the replacement of forest with a dense cover of pasture grasses reduces erosion by overland flow by significant amounts—at least an order of magnitude. However, it should also be noted that the forests in which she measured erosion rates had been disturbed by human actions, notably logging, burning and the construction of logging tracks, although the effects of these actions have not been quantified (plate 5.1).

SUBSURFACE WATER

As discussed above, the increased amount of rainwater reaching the soil surface following clearing generally results in an increase in overland flow. It is less clear whether there is also an increase in the amount of water which infiltrates into the soil. Changes at the soil surface, especially the 'packing' effect of raindrop impact, reduce porosity and therefore infiltration; increased water repellency at the soil surface is also common, although this effect disappears once the soil has become wet. On the other hand, the increased supply of water to the surface may lead to increased infiltration, possibly in localised areas where the inhibitory factors do not apply. Certainly a number of workers refer to increased recharge of groundwaters following clearing (for example, Peck and Hurle, 1973; Williamson, 1983; Allison et al., 1990).

The replacement of indigenous, perennial, deep-rooted vegetation with shallow-rooted, seasonally growing, exotic crops and pastures also means that less water is being drawn from soils by transpiration. Thus, regardless of whether infiltration increases, the net effect of vegetation clearance is to increase the amount of water moving in soils, either as throughflow or in the deeper, saturated zones. One result of this has been to bring often-saline groundwater tables closer to the surface (Peck and Williamson, 1987; Allison et al., 1990).

For example, in response to farmer concerns over waterlogging and the appearance of saline patches, the New South Wales Department of Water Resources carried out a survey near Albury in the Murray Basin. Changes in water levels and groundwater quality were monitored in more than 100 bores installed between 1945 and 1970. The results

(a)

(b)

Plate 5.1 'Litter dams' (Adamson et al., 1983) under forest, southwestern Australia. Contrary to the views of some researchers, erosion under forest is not negligible: these litter dams demonstrate that there is a net downslope movement of soil materials, probably accelerated by periodic fires which are designed to remove the protective litter layer (or 'fuel').

(c)

(d)

Plate (a) is a general view of the wandoo (eucalypt) forest; (b) the forest floor with its poor litter layer; (c) a miniature scarp being eroded by rain-wash (combined raindrop and canopy drip impact and overland flow), and (d) water-deposited sediment caught up behind an obstruction formed by leaves and twigs.

showed a change in water level ranging from a maximum increase of 57 m, a maximum decrease of 4 m, and an average rise of 0.6 m per year. More than 80 per cent of the bores recorded increases in groundwater levels (Rose, 1991).

Secondary salinisation and waterlogging

If the brackish or saline groundwaters come within the capillary fringe—say within 2 m of the soil surface (the depth varies with soil texture: silt textures are most effective for capillarity, coarse sands the least)—suction and capillary action draw the water to the surface. Evaporation removes the water, leaving the soluble salts behind. Note that evaporation has been enhanced by the removal of trees with their shading effect. The salts may accumulate over time until they reach concentrations high enough to kill plants. The soils are then said to be affected by *secondary* salinity, meaning that they have become saline as a result of human actions. Many parts of Australia are affected by *primary* salinity, notably the salt lake chains of the more arid areas. These areas were saline long before the advent of European farmers.

Where subsurface waters are not particularly saline, the hydrological changes described above may result in waterlogging, causing anaerobic conditions and structural decline in soils, increased overland flow, and reduced trafficability for machinery and stock. One significant outcome is reduced productivity (McFarlane et al., 1989).

Numerous other effects arising from alterations to natural hydrological processes are not discussed here, including the effects of reafforestation, dredging, erosion and flood control measures, and irrigation (Warner, 1984).

CULTIVATION AND SOIL PROPERTIES

The cultivation of land for agriculture, especially in the absence of good farming practices, has profound effects on the soil (Abbott et al., 1979; Clarke, 1986; Hobbs and Saunders, 1993). According to the Standing Committee report on land degradation in Australia, in terms of income forgone, soil structural decline is the most costly form of land degradation in Australia (House of Representatives . . . , 1989).

Cultivation physically pulverises and mixes surface and subsurface soil horizons and accelerates the loss of organic matter, thereby diminishing the fertility of the surface layer (but enhancing that of the deeper horizon) and breaking up root and fauna channels and subsoil nests. Allied with increased soil bulk densities due to the considerable pressures exerted on the soil, the effects are: loss of soil structure; the formation of plough pans and subsoil hardpans (McGarry, 1990:874, for example, illustrates severe root distortion with horizontal growth along compacted zones under cotton crops which were subjected to wet soil tillage); reduced water and nutrient uptake by crops; decreased permeability and increased overland flow; increased waterlogging and surface sealing; increased erosion by wind and water; impeded germination; increased susceptibility to disease, and declines in yields (by as much as one-third (Grant, 1992)).

WIND EROSION

The replacement of forests and woodlands with short crop and pasture species results in a dramatic change in resistance to wind. Trees and shrubs slow down the movement of air over the earth's surface, by friction (figure 5.4). Removal of this vegetation results in significant increases in wind speed near or at the soil surface. Soil materials then become highly susceptible to mobilisation by wind (figure 5.5). That susceptibility is further enhanced by the absence of a protective ground cover or litter layer, or by physical disturbance of the soil by ploughing (figure 5.6) or by stock grazing and walking over the surface. As with overland flow, any loss of cohesion between individual soil particles, by

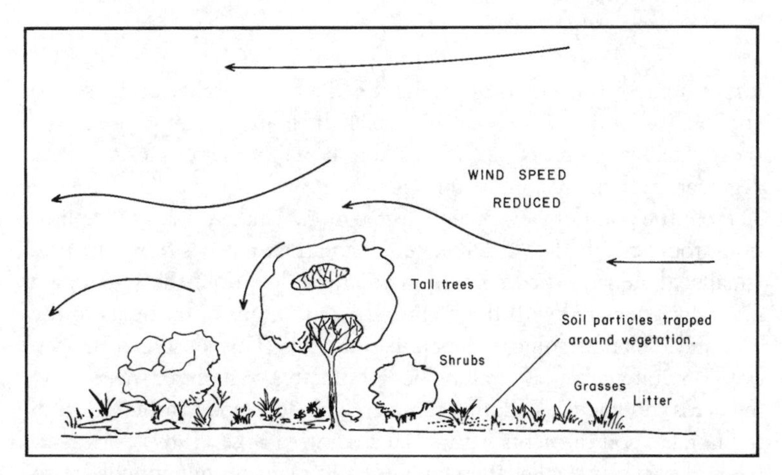

Figure 5.4 Diagrammatic illustration of the effects of vegetation on wind erosivity.

Source: Modified from a conceptual diagram by Wolfe and Nickling, 1993: fig. 1.

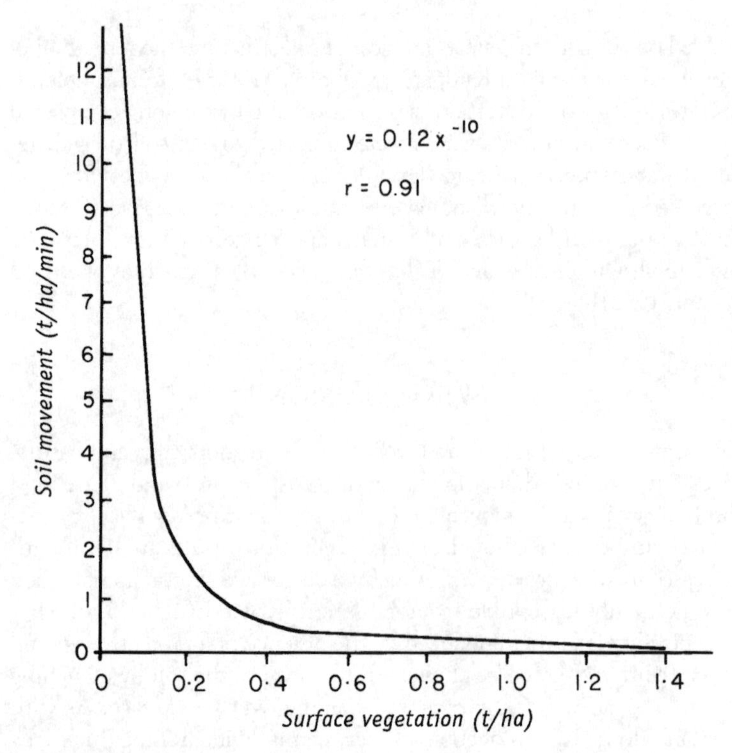

Figure 5.5 Relationship between soil movement by wind and biomass of the vegetation cover.

Source: From Scott, 1991:327.

loss of humus or soil moisture, enhances the susceptibility of the soil to erosion (Chepil and Woodruff, 1963). In many parts of Australia, accelerated wind erosion is considered to be a more serious form of land degradation than erosion by running water.

From the point of view of land degradation, wind erosion has implications from both the loss of soil materials and their redeposition. The *fines* (smaller than sand-sized particles), which are lost from the system, are generally associated with the highest concentrations of nutrients in the soil, and include significant amounts of humus. This 'dust' can be seen even on quite calm days trailing behind tractors or mobs of sheep moving across dry paddocks during summer. The dust cloud carries much of the farmer's soil nutrients with it. According to Rose (1989), for example, soil particles smaller than 0.053 mm in diameter are enriched with between 2 and 3.6 times the overall residual concentration of nitrogen, nitrate and ammonium in soils. On the other hand, redeposited sand

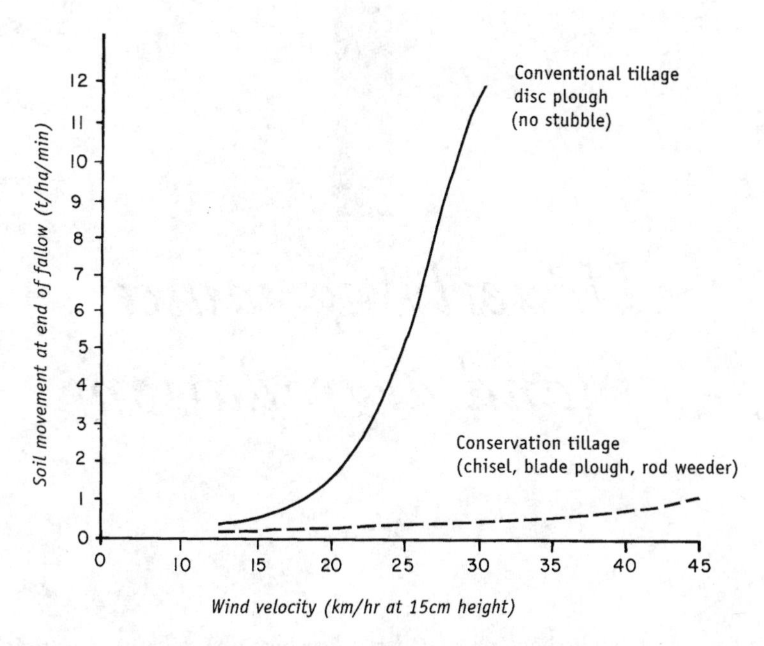

Figure 5.6 Relationship between wind erosion and tillage.
Source: Based on Scott, 1991:326.

covers fences and clogs roads, streams, dams and unprotected vegetation, and has no beneficial effects (in that it does not bring nutrients with it, just quartz). Dealing with sand drifts also incurs significant off-site costs to the community.

CONCLUSION

This chapter has considered the direct causes of land degradation. Many would appear to be reasonably self-evident; and an obvious question is: why did farmers persist with practices for so long (and in many cases still do so), when those practices were clearly damaging the environment? In order to answer this question, it is necessary to consider the more indirect, underlying causes.

6

Underlying causes of land degradation

Previous chapters have presented a somewhat depressing view of the extent and severity of land degradation in Australia. It has also been seen that the specific causes of the problems are reasonably well known. It is therefore not sufficient, if the problem of land degradation is to be resolved, to understand the immediate link between farming practices and the mechanisms responsible for (say) accelerated soil erosion or secondary salinisation. It is also necessary to ask why, if these links have long been known and understood, the harmful practices have been allowed to continue. What, in other words, are the underlying causes of land degradation?

Conacher (1978, 1986), among others, has suggested that fundamental, underlying causes of Australia's environmental problems are people's attitudes towards and perceptions of the environment; in particular, the common perception of it as a resource. To that could be added the related factors of ignorance—lack of knowledge about and interest in the biophysical environment, its extent, characteristics and response to human actions—and the recency of European settlement. It was considered that there is still a pioneering, frontier-type mentality, especially in the peripheral states and territories. That mentality views the environment as being there for human use, it is limitless, it will withstand abuse, and in any event its destruction or degradation is unimportant. Over-emphasis on economic growth was thought to maintain and encourage such attitudes, supported by an aggressive advertising industry and with multinationals playing an increasingly important role. Multinationals play a role not only through the standard examples of minerals exploration and mining, but also through high-pressure selling of farm machinery and chemicals, overseas demands for unprocessed raw materials, and powerful political influence. The latter is reinforced by potent economic and employment arguments, particularly in times of economic recession.

Government agency structures and enabling legislation reflect these realities. Departments of agriculture, forests, town planning, mines and public works are set up primarily to plan and manage resources and provide services, and only secondarily to protect and enhance the quality of the environment. Environmental legislation and agencies are often weak in comparison with development-orientated agencies. The environment agencies may proffer advice and seek public comment on development proposals, but in general they are unable to enforce environmental quality standards effectively. National parks and wildlife agencies operate under similar legislative and political constraints. Secrecy compounds these problems. Unlike developers, conservationists have little if any input in the drafting of legislation designed to permit or promote development.

Other legislative difficulties for conservationists include the deliberate avoidance of the courts in framing environmental protection acts and the inability of people to gain standing or to mount class actions with respect to environmental matters (with the significant exception, in both instances, of New South Wales).

This chapter addresses in further detail the question posed at the beginning: namely, what are the underlying factors responsible for land degradation? But at the outset, it must be appreciated that definitive answers are most unlikely to be arrived at. Rather, the chapter presents a somewhat exploratory discussion. It considers first the economic constraints and context, and the nature of the biophysical environment, in which rural land managers have to operate. This is followed by discussion of attitudes towards and perceptions of the Australian environment by early settlers, farmers and graziers, public servants and politicians, including a resistance to change and inadequate awareness of both the environment and the problems; and the effects of government agencies and policies.

ECONOMIC PRESSURES

It's hard to be green when you are stuck in the red.

A major pressure on the land base has been caused directly or indirectly by growing economic difficulties for farmers. This problem, which has been especially marked since the mid-1970s, is by no means unique to Australia (OECD, 1989; Young, 1991).

Australian farmers have been described as among the most efficient of those in developed countries, with one writer claiming Australian farmers to be 51 per cent more productive than their United States counterparts, 155 per cent more productive than British farmers, and 220 per cent more productive than German primary producers on a per capita basis (Cribb, 1989). But economic problems from the mid-1970s onwards have increasingly threatened not so much farmers' productive efficiency (annual growth rates of agricultural production since the 1960s have exceeded those for all other sectors of the economy) as their economic viability. Added to this has been a number of serious droughts around the country, one of which, in 1982, virtually halved the real net value of agricultural production (Anon., 1983).

Contemporary Australian agriculture is full of contradictions. Australia ranks among the world's major agricultural producers, exporting $14 billion worth of produce accounting for 64 per cent of its domestic production in 1990/91. But the role of agriculture in relation to the national economy has diminished steadily since the 1950s, despite big gains in the volume of production and exports, at least a doubling of crop areas (figure 6.1) and cattle numbers, and a quadrupling of the net value of agriculture (to the end of the 1980s—it has fallen since then). Together with fishing and forestry, agriculture's share of the export market dropped from 84 per cent in 1953–54 to a range of 25 per cent to 58 per cent in recent years, depending on seasonal fluctuations and export

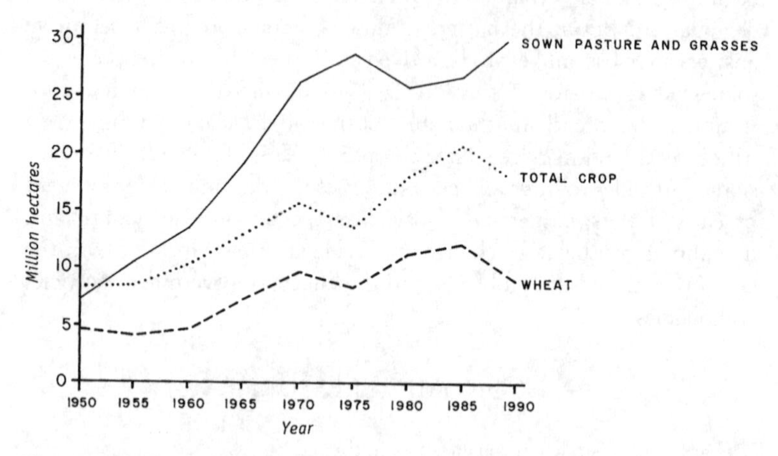

Figure 6.1 Increase in areas under crop and pasture 1950–90.
Source of data: Cribb, 1991: table 3.

demands for both farm and industrial products; and its contribution to gross domestic product plunged from 20 per cent to 4–5 per cent (Cribb, 1991; ABARE, 1992).

Although these proportional declines are due partly to the expanded role of the mining, manufacturing and tertiary sectors of the economy, accelerated declines in different components of the agricultural sector during the 1970s and 1980s were also the product of other factors. These included: global and national inflation; rising interest rates (from around 6% in the early 1970s to as high as 22% in the late 1980s (ABARE, 1992)); energy-related price increases (for example, farm diesel fuel increased from about 4 cents/litre in 1970 to 36 c/L in 1991 (ABARE, 1992; see also figure 3.3)); inability to compete with subsidised exports from competitors, notably the European Community and the United States; and slumps in agricultural commodity prices, especially for wheat, wool, dairy produce and sugar. Thus, by the early 1990s, despite actual increases in agricultural production and a levelling out or reduction of the cost of some farm inputs such as fuel and interest rates, in real terms Australia's farmers were worse off than a decade earlier (figure 6.2). According to Cribb (1991), farmers' terms of trade in 1990 were at their lowest point since the Great Depression. Over the decade to 1990, farm costs increased by an average of 100 per cent while commodity prices increased by only 53 per cent.

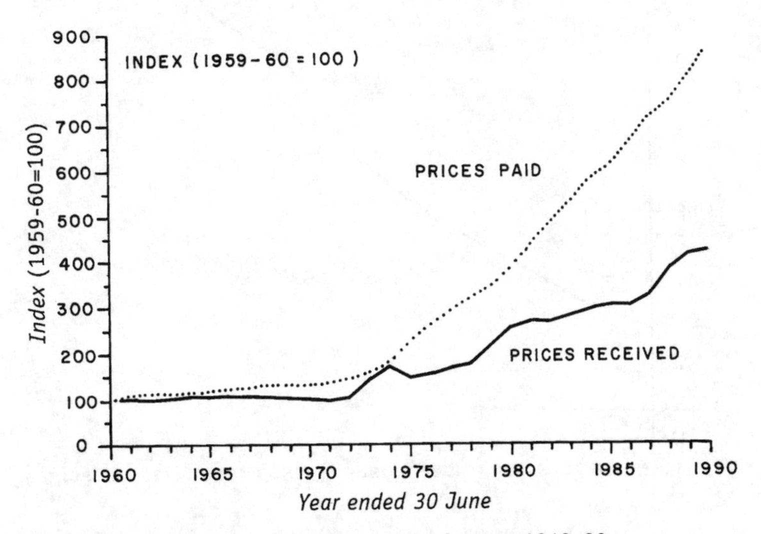

Figure 6.2 Prices received/prices paid by farmers 1960–90.

Source: Cribb, 1991: fig. 4.

As a result farm debt increased: total institutional farm debt in 1990 exceeded $14 billion (figure 6.3), averaging $84 400 per farm (Cribb, 1991). But this figure conceals wide disparities in debt burdens between farm types and regions, with the crops sector and drought-affected areas of Western Australia, Victoria, South Australia and Queensland worst affected in that year. ABARE's 1991 *Farm Surveys Report* states that broadacre cropping operations had the greatest average farm business debt at $100 000, compared with $60 000–80 000 for beef, dairy and horticulture. Only about one-quarter of all farms were without an appreciable debt. The seriousness of this situation was underlined by the fact that in 1989/90, one in six Australian farms experienced a negative income, and one-quarter had incomes of less than $5000 (Cribb, 1991). Since then, serious difficulties with export commodity prices have done little to alleviate this farm crisis.

Farmers have responded to the increasing cost/price pressures in a number of ways, producing some important structural changes in the

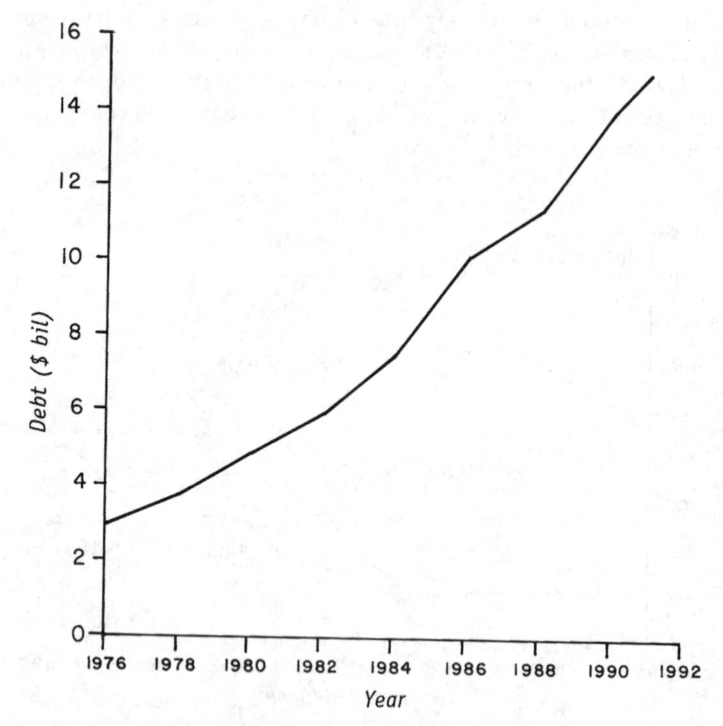

Figure 6.3 Total institutional farm debt 1976–91.
Source of data: ABARE, 1992: table 29.

agricultural sector with both direct and indirect implications for land degradation. Although increased capitalisation, mechanisation, labour reduction and increased reliance on the family unit were already well-established trends, during the decade of the mid-1970s to the mid-1980s, there was a marked swing to increased farm sizes and areas under crop. The number of farms decreased from 205 000 in the early 1950s to around 165 000 in the late 1980s (ignoring the change in definition of 'farm', or rural establishment; see figure 6.4). This change was accompanied by a loss of between 70 000 and 100 000 from the rural workforce (figure 6.5), despite a significant increase in Australia's total population over the same period.

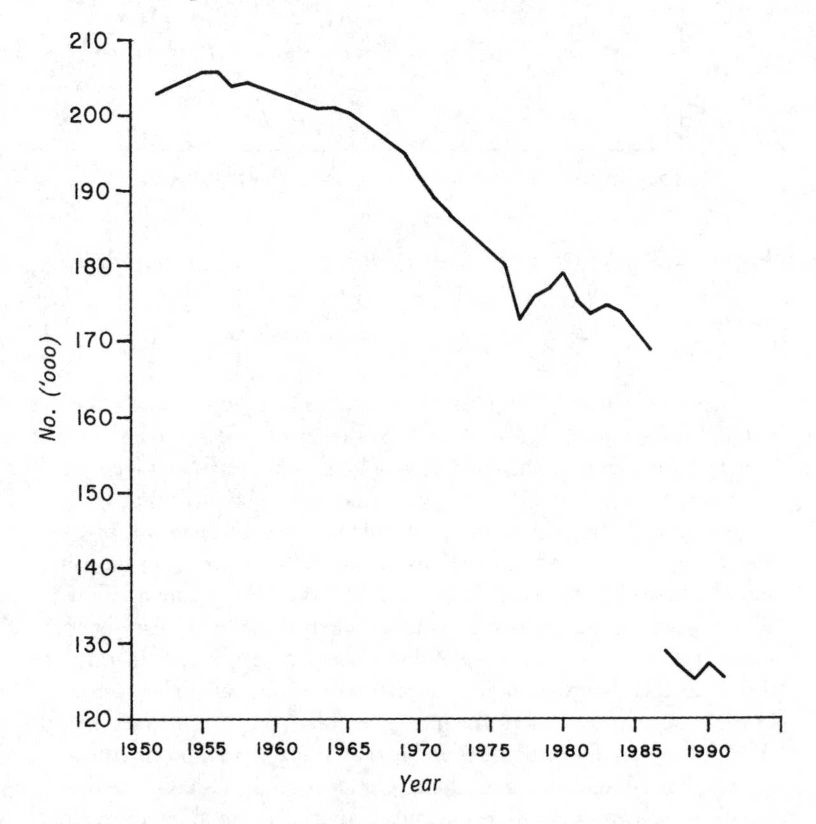

Note: from 1986/87, only rural establishments with an Estimated Value of Annual Operations (EVAO) of $20 000 or more were included, whereas prior to that date, EVAO was $5000 or more.

Figure 6.4 Australia: number of agricultural establishments 1952–92.
Source of data: ABARE, 1992: table 24.

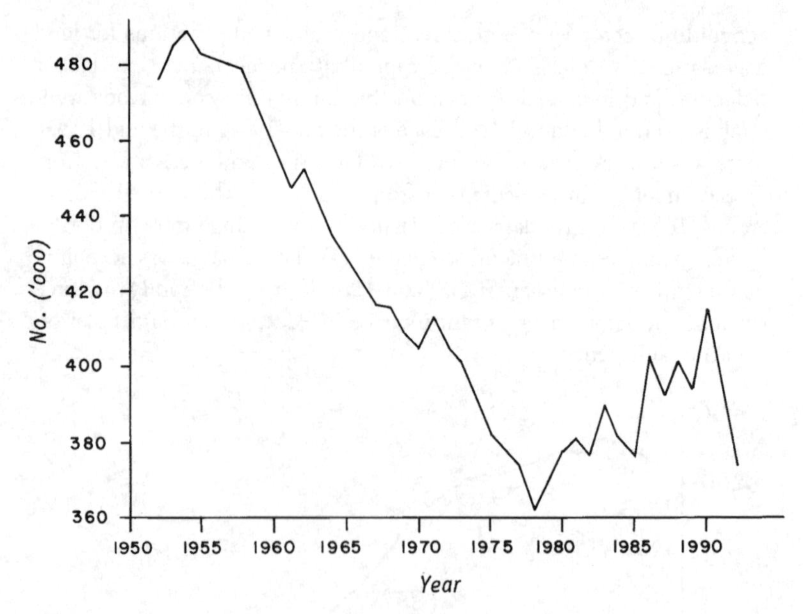

Figure 6.5 Australian farm employment, including employers, self-employed, wage and salary earners and unpaid family helpers.
Source of data: ABARE, 1992: table 24.

Associated with this has been a dramatic increase in the area per agricultural worker, according to Australian Bureau of Statistics data used in a practical exercise for first-year geography students at the University of Western Australia in 1994. In 1933, the area per agricultural worker ranged from 276 to 653 ha in 36 wheatbelt shires. In 1986 this had increased to a range of 451 to 2639 ha per agricultural worker in the same shires. These changes have obvious implications for the care and intensity with which land is farmed; and although these or similar data are sometimes quoted as a measure of the Australian farmer's efficiency, this is not necessarily an accurate interpretation of the data. They can also indicate huge areas of land which may be visited only a few times during the year, weed-infested pastures and broken-down farm infrastructure.

Another notable feature of the cost/price squeeze has been the deferral of farm expenditures in several areas, especially capital improvements (Knopke and Harris, 1991). These deferrals may have important consequences for the management of land degradation. Figure 6.6, for example, illustrates the overall decline in tractor sales, which lifted in 1988 but dropped markedly again two years later.

Significantly, an Australian National Opinion Polls (ANOP) survey

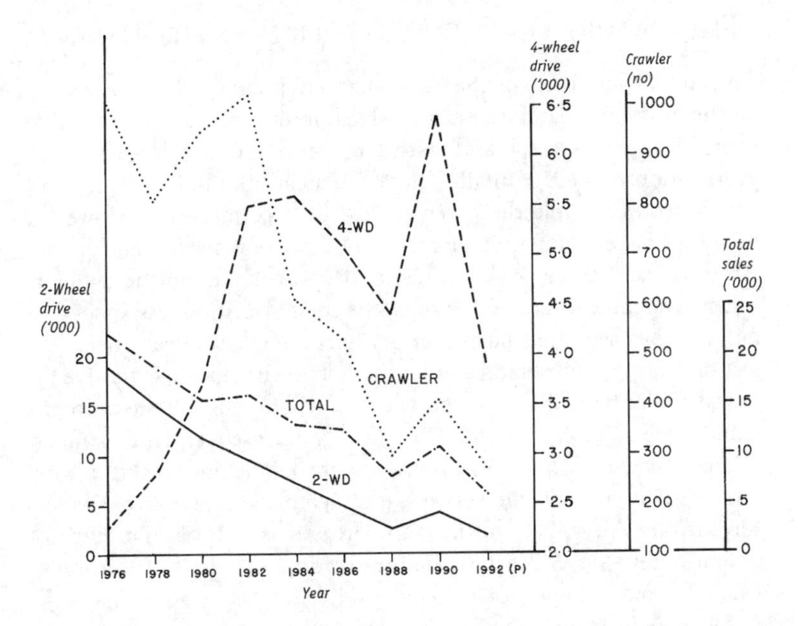

(P) = provisional figure.
Note: data before 1979/80 not comparable due to revision to series.

Figure 6.6 Numbers of new tractors sold in Australia, by type, 1976–92.

Source of data: ABARE, 1992: table 21.

for the National Farmers' Federation in 1991 found that 53 per cent of the farmers surveyed planned to reduce expenditure on Landcare and conservation activities due to their changed income expectations. A further 9 per cent were undecided, and about one-third planned to stop their expenditure altogether. The activities which would not be carried out included tree planting, soil and water conservation works, and fencing (Farley, 1991). An ABARE 1991/92 Landcare and drought management survey underlined this trend. In this case, the vast majority of broadacre farmers considered that cash availability was a constraint in implementing land degradation solutions, although this feeling was less strong for wheat than other crop industries (Nelson and Mues, 1993).

Finally, these agricultural shifts frequently involved greater use of marginal lands and tighter crop rotations (including continuous cropping), the use of bigger and more powerful agricultural machinery (note the relative sales of two- and four-wheel drive tractors in figure 6.6), and heavier use of agricultural chemicals (table 3.1 and figures 3.2 and 3.4) in attempts to maximise productivity and to stay ahead of costs, with important consequences for the environment (chapter 3).

THE NATURE OF THE AUSTRALIAN ENVIRONMENT

Several characteristics of the Australian environment have a bearing on the problem of land degradation. First, most of the continent (more than 70%) is arid or sub-arid, with long periods during the year when evaporation exceeds rainfall. These dry areas are also relatively flat, which makes possible the extensive use of large machines. However, their soils have been forming for many millions of years: for much of dry Australia, and in marked contrast with most of the northern hemisphere, the most recent period of continental glaciation was about 250 million years ago. Soils have been forming unimpeded ever since, subject of course to fluvial and wind erosion, but without the extensive removal and/or burial by ice sheets which ended in much of the northern hemisphere as recently as 10 000 years ago. Consequently, Australian soils are leached of most soil nutrients: their inherent fertility is very low. In addition, there are very few fresh, permanent or seasonal waterways in dry Australia, and in many areas it is only the presence of groundwater—often brackish, saline or irreplaceable 'fossil' water—which has made farming of any kind possible.

Second, the better-watered areas are located along the eastern, southeastern and southwestern edges of the continent, and it is here where most of the population lives. Unfortunately, from the agricultural point of view, these are also the areas with the most intensive land-use competition. In addition, large areas are relatively steeply dissected and consequently have thin, erodible, skeletal soils; the most fertile areas are often located on limited areas of alluvial sediments adjacent to the larger rivers. Higher rainfall combines with soil age to produce leached, naturally acidic and hard-setting soils in many areas.

A third feature of the Australian environment was its unfamiliar vegetation to the early settlers. Despite their height and density in forested areas, trees are not always a reliable indicator of soil fertility, contrary to some of the early expectations (Cameron, 1977; Bligh, 1980). Dashed hopes of a 'paradise on earth' were poignantly expressed in the words of one early settler: 'the man who reported this land good deserves hanging nine times over' (Eliza Shaw, *Swan River Colony 1830*, cited in De Garis, 1979:1). This point is expanded on by Bolton (1981) and in the chapters by Jeans and Powell in Heathcote (1988).

One of the most unfortunate consequences of the struggle to adjust to the strange environment was the introduction by the European settlers of more familiar plant and animal species, many of which apparently found the environment to their liking and developed to plague proportions (cf. chapter 2). These attitudes persist. Many recent immigrants, and some local authorities, consider the 'gum tree' to be untidy. They prefer to plant introduced, deciduous, garden and street trees—despite the need of these plants for frequent watering and their inability to provide an ecological niche for native fauna.

PEOPLE'S ATTITUDES

As implied above, people's attitudes have had an important influence on the direct and indirect causes of land degradation in Australia. On the one hand, there was an alien environment which the (predominantly) Europeans, who came with 'visions of hope', farmed using crops, animals and cultural practices which were familiar to them. But they had no experience of the climatic extremes of drought and flood, or of fire, or the unproductive, hard-setting soils.

On the other hand there was (and to a certain degree still is) the Europeans' Judaeo–Christian attitude to the biophysical environment. In essence, this attitude is the anthropocentric ethic of human dominance of nature, revealed in the Book of Genesis: 'God having created Adam, gave him dominion over every living thing' (O'Riordan, 1976:203–8; see also Houston, 1978). Land is seen as an 'asset' or 'resource' to be exploited or used for human benefit. In the present context, it means that farmers attempt to maximise production and profits and minimise costs, rather than *optimising* productivity in balance with parallel objectives of maintaining wildlife, improving soil fertility and water quality, and preventing erosion. Because they are seen as 'unproductive', residual stands of vegetation are removed (plate 6.1).

Australian society assesses the success of farmers by the sizes of their properties, the numbers of stock and machinery they own, the values they obtain for their produce at auction, and their incomes as reflected in their visible lifestyles (in this latter respect they are no different from their urban counterparts). Agricultural agencies, stock agents, banks, accountants, farm advisers and manufacturers of farm implements all encourage this approach. The 'dominance over nature' attitude has been reinforced by advertising which urges the potential buyer of farm machinery to 'tame this sunburnt country' (*Western Farmer and Grazier*, 12 May 1983). Seen in this light, it is no accident that environmental protection or conservation agencies have been so ineffectual in rural areas. Their explicit objectives—to maintain and improve environmental quality for its own sake—have been given a low priority.

However, there are strong indications that these attitudes started to change during the 1980s, if not earlier (chapter 7). In 1982, for example,

Plate 6.1 Treeless paddock, Western Australian wheatbelt. This photograph well illustrates the wholesale clearing by farmers in the name of operational efficiency. The plate also illustrates the effects of a severe drought, since the scene was photographed near Merredin in August 1980—a time of the year when the wheat crop would normally be approaching knee height. The soil is totally unprotected from the next rain- or windstorm.

Newman and Cameron surveyed 1529 people in Western Australia, grouping their findings into people from the five richest and five poorest metropolitan Perth suburbs, and seven country districts. Despite these differences in location and income, the three groups were found to have a great deal in common, with all sharing a concern for the environment. An impressive 96 per cent of the respondents felt that we should be doing more to look after the environment. Of 45 issues listed in the questionnaire, 19 were identified as being significant problems to the respondents, scoring 2.5 or more in table 6.1. What is interesting is that nine of those significant issues were predominantly rural-based problems.

Several other surveys conducted during the 1980s and early 1990s also found strong support for environmental issues, with soil conservation generally ranking high in lists of priorities (Eckersley 1989). An Australian National Opinion Poll (ANOP) survey conducted in 1991 found that economic issues, particularly unemployment and the state of the economy (the recession), dominated the community's agenda. But

> the environment now occupies a permanent place on the national agenda . . . While unemployment clearly overshadows the environment as a *current* issue, the environment emerges as a more important *long-term* issue. When Australians think ahead a decade, it is the environment which emerges at the top of their crystal ball agenda. (ANOP, 1992:6)

The community's top priorities for additional government attention were industrial waste disposal, water and air pollution, land degradation, recycling, ozone depletion and the greenhouse effect (ANOP, 1992:8)—which makes an interesting comparison with the priorities listed in table 6.1. In New South Wales, Dragovich (1990, 1991) found that urban dwellers were prepared to participate in the repair of degraded land, with approximately 80 per cent of those surveyed willing to pay a soil conservation surcharge on bread. And at an international level, Dunlap et al. (1993) found that people in the world's poorer

Table 6.1 List of issues considered by the populations (5 rich Perth suburbs, 5 poor suburbs, 5 country areas—total 1 529 people) surveyed by Newman and Cameron, in order of importance.

Issues	*Total survey*	*5 Rich subs*	*5 Poor subs*	*Country*
Nuclear war	3.6	3.2		
Dieback in forests	3.4		3.1	
Cutting forests faster than they grow back	3.3			
Loss of rare native plants and animals	3.2	3.0	3.4	3.5
Industrial pollution	3.2	2.9	3.4	
Radioactive waste from mining	3.1			
Lack of food in poorer countries	3.1		3.4	
Soil erosion on farms	3.1		2.5	
Increasing salinity of rivers	3.1	2.9	2.8	3.4
World population growth	3.0			
Increasing salt land on farms	2.9		2.3	3.3
Spraying chemicals for pest and weed control	2.7	2.4		
Wasting energy	2.7		2.9	
Declining reserves of oil	2.6			
Garbage disposal without recycling	2.6	2.4	2.4	3.1
Nuclear power	2.6	2.4	2.3	2.9
Overgrazing of agricultural land	2.6	2.8	2.4	
Vehicle exhaust emissions	2.5			
Loss of wetlands	2.5		2.3	
Airborne dust in workplace	2.4	1.8	2.6	
Bauxite mining	2.3			
Inadequate public transport	2.3	2.1		
Noise	2.2			
Lack of parks and reserves	2.1	1.9		2.4
Urban sprawl	1.9	1.7		
Freeways	1.7	1.3		
Average value	2.7	2.6	2.7	2.8

Note: a further 17 'Other issues' are not included here. Average values are based on assigning the severity of the issue from '4'— 'a very pressing problem', to '1'— 'not really a problem', and '0'— 'don't understand'. Data in the 'Suburbs' and 'Country' columns are shown only where they diverge by more than 0.1 from the 'Total survey' data.

Source: From Newman and Cameron, 1982: tables 1 and 2.

countries are just as concerned over the state of the environment as those in the wealthier ones. The survey of 24 nations showed that while specific problems varied from country to country, there was a high level of citizen awareness of environmental deterioration and support for environmental protection.

PIONEER ATTITUDES AND THE RURAL ETHOS

The relative recency of European settlement in Australia—from a maximum of a little over two hundred years to less than fifty years—is closely associated with the 'domination of nature' attitudes discussed above; and it is another of the more important factors responsible for land degradation (Holmes, 1981a, 1981b). In some rural areas there is still a pioneering, frontier mentality or outlook (Holmes, 1988), where settlement is still in the development phase, especially in parts of Western Australia, Queensland and the Northern Territory. In general terms, it is mainly in New South Wales and Victoria where settlement has passed from the pioneering to a maintenance or consolidation phase. Such regional differences reflect the different durations of European settlement and they undoubtedly have a bearing on the greater expenditure effort being devoted to rural land management in the latter states than in the former regions. Yet even in southeastern Australia, the term 'conservation' is defined and used in a utilitarian sense to mean 'wise use' (Costin and Frith, 1974; see also Conacher, 1986:317).

Contributing to the frontier mentality was the nature of the early settlers themselves. A harsh environment characterised by heat, drought dust, isolation, flood, fire and flies could be settled successfully only by people who could adjust to such conditions. This environment engendered a hard, rugged individualism and egalitarianism (Heathcote, 1977)—the 'battlers'—who did not take kindly to bureaucratic regulations imposed (often from considerable distances) by the bourgeoisie. Even today, the pages of rural newspapers perpetuate this rural-urban dichotomy, while rural events are largely ignored by the metropolitan media, at least until some disaster (read: 'newsworthy event') occurs.

This distinctive rural ethos both resulted in and was influenced by a strong, conservative, rural-based political party, reinforced by effective pastoral and agricultural pressure groups, and strengthened in the 1980s by a revitalised National Farmers' Federation. Partly as a result of imbalanced electorates which gave rural voters considerably more influence than their urban counterparts (justified initially by the huge areas of some of the electorates), and partly due to the strong export orientation of Australia's primary industries—and their major contribution to export earnings—country people have long had a very strong influence on the composition and policies of Australian governments (McDonald, 1976). Interwoven in all this has been the notion that farmers are the 'backbone of the country', and that society is indebted to them for the availability of cheap and abundant food and a high standard of living. However, that influence diminished considerably in the late twentieth century.

FARMER EDUCATION AND RESISTANCE TO CHANGE

One problem encountered in dealing with land degradation in Australia has been farmer resistance to change, especially with respect to better management practices. There are a number of possible reasons for this, many embedded in rural sociology. Dealing with these barriers is important in order to address land degradation problems. As discussed above, farmers are independent people who, as private landowners, see themselves as having the right to make decisions as they see fit with regard to their own land. They often resent external expert advice which may be seen as interfering and paternalistic. In addition, social structures and networks in rural communities strongly affect individual, family and group behaviour. Few wish to be seen as non-conforming or critical for fear of loss of acceptance by the community. Factors such as poor understanding of issues (often based on inadequate information), resistance to risk taking, lack of incentive, ingrained habits, conservatism, and several generations on the land, all influence the effective transfer of information, acceptance of new ideas and the modification of behaviour (examples are provided in Rickson et al., 1987; Hollick, 1990; Vanclay, 1992). Fortunately, much of this is changing rapidly and some reasons for that are discussed further in chapter 7.

One barrier to change has been the relatively low level of education among farmers, especially at advanced secondary and tertiary levels (Sexton, 1980), and this frequently had a bearing on attitudes to land degradation and willingness to change. Even at the end of the 1980s, only about 25 per cent of the farm workforce had school leaving, trade or tertiary qualifications. Although this was a significant increase on previous decades, it is a low figure when compared with 50 per cent of New Zealand farmers and 90 per cent of farmers in western Europe with equivalent qualifications (Cribb, 1991). According to Campbell (1992), less than 10 per cent of country children go on to tertiary studies, compared with nearly a third of their urban counterparts. This situation is exacerbated by the increasing number of rural school closures and smaller country centres (see chapter 4).

In combination with the factor of isolation, the relatively basic level of education meant that farmers applied methods which were either known to them or their fathers, or which they had tested by trial and error. For example, Williams (1980) cited a Queensland study which noted that farmers in the severely eroded sugarcane growing areas had been urged since 1941 to implement soil conservation measures. Yet by 1974, only 2 per cent of land had been protected. In Western Australia, A.J. and J.L. Conacher (1988) reported that in the early 1970s, soil conservation advisers considered that the major impediment to the implementation of standard soil conservation practices was the farmers' failure to recognise either the fact of soil erosion or its significance, although this could also indicate agency failure to communicate effectively. More recently, an Agrimark survey cited by Rudwick (1991) found that 70 per cent of farmers saw themselves as being conservative and traditional. For example, only 13 per cent would allow new techniques or products to be tested on their properties, and over half (wisely) said they would not use new chemicals until they had been well tested in the district.

LACK OF AWARENESS

Unfortunately, for some of the reasons given above, there has been a disturbing degree of ignorance among many farmers of their fundamental resource—the physical environment. This was demonstrated by an unpublished survey carried out in 1984 in the Western Australian wheatbelt by Lobry de Bruyn and Mackenzie (discussed in A.J. and J.L. Conacher, 1988). Several examples can be cited.

It was found that drought frequency was underestimated and viewed incorrectly as a cyclic phenomenon. One farmer on the eastern margin of the wheatbelt reported a mean annual rainfall more than double the actual value of under 300 mm. Although virtually all the farmers considered secondary soil salinity to be a problem, very few saw it as being a consequence of the removal of the native vegetation, the fundamental cause. Despite the considerable importance of organic matter in relation to soil fertility, moisture retention, soil biota and soil structure, 33 of the 41 farmers had no idea of their soils' organic matter content and only 4 of the remaining 8 farmers gave a figure that approximated the actual value. Soil acidification is an increasing problem in many areas, yet 13 farmers did not know the pH of their soil, 18 did not know at what pH level acidity becomes a problem, and 20 did not know whether their pH was changing (having had their soils tested only once). Many acknowledged that some of the land degradation problems were the result of overcropping and overstocking; yet 29 of the 41 farmers had changed their farm rotation practices for economic and not environmental reasons.

One decade on, this lack of knowledge by farmers of their soils is still apparent. According to a Commonwealth Landcare survey in 1992/93 of over 1800 farmers, of those who possessed farm plans (and who are therefore presumably reasonably motivated), approximately 80 per cent of broadacre farmers and 90 per cent of dairy farmers did not have information on the soils or land capability on their plans (Mues et al., 1994).

Despite some reservations expressed below, this somewhat alarming level of ignorance was perpetuated by the overwhelming tendency of Western Australian wheatbelt farmers to obtain their advice from other farmers in preference to the state agricultural agency. Fry and Goss (1985:73–4) found that although the amount of soil conservation activity by central wheatbelt farmers was 'high', 78 per cent of their soil conservation links were with other farmers and only 5 per cent with Department of Agriculture officers.

It is not known whether the situation is comparable with farmers from other parts of Australia; but it should be noted that this situation is changing with Land Conservation Districts and Landcare (chapter 7) and active work by agencies to improve their extension services.

GOVERNMENT AGENCIES

An important factor contributing to the context in which land degradation developed was and is the nature of the government agricultural agencies. As is true of most institutions, they have rigidly hierarchical structures. Decisions are made at the top and transmitted down; information flows from the bottom up are repressed by the nature of the institution itself (and, it must be added, by human nature). In addition, personnel are promoted largely on the basis of seniority, which means that many of those who are now in a position to make and implement decisions first obtained their qualifications many years ago.

Moreover, at least in some states, the tertiary education of agricultural agency officers included little training in the nature and causes of soil erosion in particular and land degradation in general. Even as recently as 1985, very few agricultural faculties in Australian tertiary educational institutions listed courses in land management (Anon., 1985). They were strong in agronomy, animal husbandry, agricultural economics and, to a lesser extent, farm machinery, fertilisers and chemical methods of pest and weed control, but deficient in the processes and mechanisms involved in land degradation and hence the appropriate management responses. As a result, agency officers learned how to manage land degradation 'on the job', under an older officer, before being placed in charge of their own districts. Existing methods were perpetuated and, inevitably, mistakes were made.

One example occurred in a soil conservation pamphlet distributed by a Western Australian Department of Agriculture officer around 1970 (reported in A.J. and J.L. Conacher, 1988). The pamphlet included a diagram illustrating a system of grade banks. They were depicted as diverting overland flow to grassed waterways which then discharged the water on to salt-affected land. Farmers who applied the agency's advice in the pamphlet may have controlled soil erosion on their paddocks, but quite possibly with three adverse consequences: 1) wasting precious water in this seasonally arid environment; 2) initiating accelerated erosion along the 'grassed' waterways, since a grass cover is virtually impossible to maintain over summer, and 3) exacerbating the salinity problem, since the fundamental requirement in saltland rehabilitation is to dewater the salt-affected area. We stress that this example is included to illustrate the kinds of errors which were made in the past.

Although soil conservation legislation has been adopted by all Australian states, except Tasmania, from as early as the late 1930s (New South Wales and South Australia) to 1969 (the Northern Territory), the agencies and soil conservation practices were largely modelled on the United States Department of Agriculture's Soil Conservation Service (Bradsen, 1988, chapter 2): but note that Western Australia had a *Sand Drift Act* as early as 1889, and South Australia in 1923. Significantly, it was only in 1971 that the first Australian soil conservation conference was held (in Melbourne). Participants (including one of the authors) were struck by the previous lack of communication among the various agencies and personnel, the lack of knowledge concerning common problems and the means of dealing with them, and the duplication of effort that was taking place.

Only in the 1980s did this situation start to change to any significant extent as far as the quality of tertiary education is concerned. But agricultural agencies and research institutions are still *resource*-orientated; that is, their 'charter' is still first and foremost efficiency in production, and they do not consider it their responsibility to maintain and improve environmental quality *per se* (see further, Reeve et al., 1990). The agencies have been encouraged in this by societal attitudes as expressed through the terms of the agencies' enabling legislation (again, this is changing), by farmer pressure groups, by the rural media, and—by no means least—by multinational companies.

Where existing legislation does give agricultural or soil conservation agencies powers to intervene and enforce ameliorative measures, this power is very rarely used (for good reason—agricultural extension officers rightly rely on co-operation with farmers). But in certain instances, especially in pastoral, leasehold areas, some of the failures to enforce the provisions of those leases seem to be inexcusable. The political power of the rural lobby referred to earlier is not irrelevant here (L. and B. Chatterton, 1982).

GOVERNMENT POLICIES

As discussed by the Conachers (1988), several government policies also contribute to land degradation, albeit unwittingly. Referring to Blyth and Kirby from the (then) Bureau of Agricultural Economics (Anon., 1984), the Conachers noted that the leasehold system, especially but not exclusively in the extensive pastoral areas, has provided little incentive for landholders to maintain or improve the quality of the land and its vegetative cover, although theoretically it has been easier to enforce management changes on leasehold than on privately owned land. This is exacerbated by the often uncertain duration and security of tenure.

Fertiliser subsidies may have encouraged the substitution of fertilisers for soil nutrients which might be retained by better land husbandry, and may also have encouraged agricultural expansion into more marginal, erosion-prone lands as well as aggravating problems such as eutrophication of waterways. This subsidy was removed in 1988.

The artificially low price of irrigation water has led to inefficient irrigation technology and wastage of an increasingly scarce resource, and has probably tended to exacerbate salinisation and soil structural decline in irrigated areas. Price increases met with angry grower resistance in Victoria. Drought relief programs have arguably resulted in sound land management being discarded for subsidised fodder, subsidised freight for water, feed and livestock and concessional finance. Retaining livestock during a drought results in accelerated pasture and land degradation. Without fodder and other subsidies and relief programs, it is likely that only a core of breeding stock would be retained, with less severe impacts on the drought-stressed environment. These policies have been reviewed (DPIE/ASCC, 1988). Tax concessions for clearing land were also indirectly responsible for much land degradation, since they encouraged unnecessary vegetation clearance; those concessions were removed in 1983.

Similarly, price supports for crops, and export policies, may have been partly responsible for the over-intensive use of land. Even subsidies for soil conservation work could lead to the development of erosion-prone marginal land or its transformation from grazing to crop production. While subsidies may increase soil conservation efforts, they may also increase the need for those efforts, particularly in the case of less productive farms.

Nevertheless, it should be recognised that by and large Australian farmers are far less subsidised than their northern hemisphere counterparts, although special relief programs such as drought and other relief are usually not built into calculations. United States Department of Agriculture estimates reported by Cribb (1989:30) were that the 'producer subsidy equivalent' for Australia was 9 per cent in 1982–84, compared with 22 per cent in both Canada and the United States, 33 per cent in the European Community and 72 per cent in Japan.

A string of grandiose schemes dreamt up by various politicians (and agricultural scientists) also warrants a mention. The schemes were encapsulated in phrases such as 'watering the desert', clearing a 'million acres a year', and 'populate the north', reflecting a desire to settle the empty regions of Australia. Some have been mentioned in chapter 4. But the ideas of securing 'untapped potential' invariably foundered due to the poor understanding of the fragile and inhospitable environments involved (incurring problems such as massive erosion of tropical soils, siltation of dams, huge inputs of pesticides to cope with unexpected pests, loss of crops due to birds, and the potential for increased diseases), as well as fundamental economic problems such as isolation from markets and poor returns on the massive injections of funds (Courtenay, 1988).

Conclusion

The causes of land degradation involve both the nature of the biophysical environment and the human responses to it. Essentially, 'the environment' has been regarded as a resource to be manipulated for economic benefit. However, the methods used for this purpose increasingly are being seen to be causing severe damage, thus threatening the long-term viability of the environment viewed both for its own sake and as a resource. Any improvement requires attitudinal changes by farmers, agencies and governments.

There are indications that attitudes to the environment have changed quite dramatically, especially over the past decade, but even so, a great deal needs to be done. Fundamental additional constraints are institutional structures, lack of funds, and insufficient knowledge both of the problems themselves and of the best rehabilitative measures to use. While there is room for some optimism, a considerable degree of inertia in educational institutions, agricultural agencies and among farmers and politicians still needs to be overcome.

The previous chapters of this book outlined the nature and extent of the problem of land degradation in Australia, and drew attention to the problem of data availability and reliability. Land degradation has consequences and implications which extend well into the future and beyond the farm gate: it is a major, national (and international of course—Australia is by no means unique in this regard) problem for which solutions must be sought and implemented. Direct causes of land degradation were discussed in chapter 5 and some of the more subtle, underlying causes were considered in the present chapter, in an attempt to answer the question: why, if the problem and its immediate causes are reasonably apparent and have been known for such a long time, does land degradation continue, evidently with increasing severity? The next chapter considers some possible solutions; but it needs to be emphasised that this is not an instruction manual for farmers seeking to solve their waterlogging, acidity, hardpan, water repellency, nutrient depletion, weed or insect pest problems. For that, and despite the reservations expressed earlier in this chapter, recourse should be had to the nearest state government agricultural or soil conservation agency or its equivalent.

■

7

Solutions to land degradation problems

For sustainable agriculture to work, farmers must be given options that do not threaten the economic viability of their farms yet encourage them to become as good managers of the environment as they are of food production.

(Government of Canada, 1991:9.27)

Solutions to the problem of land degradation and its wider implications are required both on and off the farm. There are specific things that farmers can and should do, and there are other things that can only be done by the broader local, regional, state or even national communities. This chapter tends to focus on the farm or local level, although actions at broader levels are also discussed. The approach taken is to recall the types of land degradation problems and then to consider how these may be or are being dealt with.

As indicated at the end of the previous chapter, what is presented here is a discussion of various possibilities, not a soil conservation 'cookbook', for which there are many excellent texts, including Hudson (1981) and Morgan (1986), as well as the compendium edited by El-Swaify et al. (1985), among other publications of the Soil Conservation Society of America. Readers are also referred to the Soil and Water Conservation Association of Australia (PO Box W21, West Pennant Hills, NSW 2125) and to the World Association of Soil and Water Conservation (317 Marvin Avenue, Volga, South Dakota, USA 57071). For management specific to Australian land degradation problems, see also appendix A of Western Australia's *Decade of Landcare*

(Butcher, 1992), the *Landcare Manual* by Roberts (1992) and the *Australian Journal of Soil and Water Conservation.*

LOSS OF HABITAT AND PROTECTING ECOSYSTEMS

The first, most generally recognised solution to the problem of loss of habitat is to create an adequate system of reserves. For Australia as a whole, ideally every ecosystem (terrestrial, aquatic and marine) should be represented. Table 7.1 shows the areas currently set aside as conservation reserves of all kinds. There is obvious difficulty in reserving areas of those ecosystems which have already been destroyed or severely altered. Another problem is the competition for increasingly scarce land and other resources, where the value of ecological reserves is difficult to defend (in a predominantly materialistic society) against the monetary and employment values derived from 'development' and exploitation of timber, soil, water or mineral resources.

Other issues concern the size, shape, spacing and connectivity of reserves. For ecological purposes, in general the more compact the reserve the better, as this minimises the 'edge effect' whereby adjacent land uses impinge on the reserve's ecological integrity. Reserves should be close enough to one another, or connected by corridors, to permit the movement of wildlife (figure 7.1). Their size should be sufficient to allow the genetic diversity of populations to be maintained. However, in nearly all cases there is inadequate scientific knowledge available to

Table 7.1 **A summary of the number, absolute and proportional areas of nature conservation reserves in the various states and territories and in total in 1988. This overall category includes all reserves, parks and other lands set aside for conservation landuse.**

State	*Number*	*Area (km²)*	*Proportion of state (%)*
NSW	876	46240	6
Vic.	724	36530	16
QLD	1148	73280	4
SA	558	222340	23
WA	2494	305040	12
Tas.	454	19340	29
NT	180	80230	6
ACT	12	2240	93
Australia	6446	785240	10

Note: the inclusion in these data of forestry reserves, which were not created for conservation land use, inflates the figures. There is also a large number of very extensive reserves in arid regions.

Source: Graetz et al., 1992: 131.

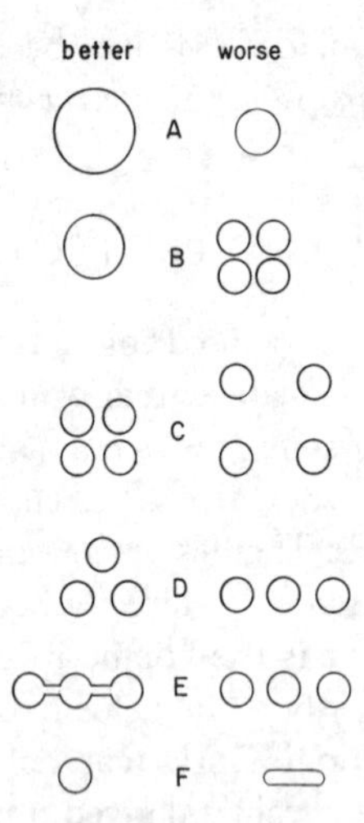

Figure 7.1 Geometric principles for the design of nature reserves.
Source: Diamond, 1975, in Gell and Mueck, 1987.

determine with confidence what these minimum sizes are for various species. Island biogeography provides some indications of answers to these and other issues (Abbott and Black, 1987; Diamond, 1975; Gell and Mueck, 1987; MacArthur and Wilson, 1967).

In 1988, 10 per cent of the Australian land mass was set aside for conservation land use (but note the range of percentages in the various states and the qualifications in the caption to table 7.1). In addition to questions concerning the adequacy and representativeness of these reserves, the question arises as to what should be done with the 90 per cent that is not reserved. Should people be allowed to deal with it in any way they wish? Few would agree with such a proposition—for many reasons. These include: the damage caused to other people both spatially and through time (future generations) by the selfish actions of a relative few; ethical considerations; and spiritual arguments, such as the proposition that human beings have stewardship over other forms of life and are under an obligation to protect those other forms which are unable to protect themselves.

As previously discussed, most habitat loss has been caused by clearing native vegetation for agriculture, which covers only some 8 per cent of the country. Thus it follows that attempts to preserve flora and fauna outside any system of reserves should focus on the pastoral (rangelands) and other areas outside those used for intensive and extensive cropping and grazing on improved pastures. Though variably degraded, these areas still retain their original vegetative species, and protection from further degradation should be relatively straightforward. The difficulties have more to do with political, economic and social realities than with technical problems.

In contrast, rehabilitation of ecosystems which have been totally lost, or are in serious danger of being destroyed, has to take place in agricultural areas (see further, Hobbs and Saunders, 1993). Farmers need to be encouraged to protect areas of remnant vegetation and to replant other areas. The values of doing so are many and include: the conservation of flora and fauna (including preservation of species which may be of economic importance in the future); provision of habitat (for birds, insects, animals and micro-organisms, many of which are important as natural predators, parasites, pollinators and nutrient cyclers); water and salinity management; protection from erosion by water and wind; micro-climate modification (temperature, moisture and wind, reducing stress on animals and plants); greenhouse gas reduction; and recreational, aesthetic and other economic uses (for fuel, honey, timber, seeds, medicines and pesticides).

Management options available to farmers for these purposes include: protection from grazing (fencing off); replanting (using local species, including understorey); providing shelter for wildlife (including ensuring presence of an adequate litter layer, rocks and hollow logs); providing linking corridors between patches of bush; management of drift from chemicals (careful application), and control of exotic weeds and feral animals (strategic use of herbicides, baiting and traps). Books by Breckwoldt (1983) and Hussey and Wallace (1993) develop these options further.

Davidson and Davidson (1992) have discussed the importance of bushland ecosystems in agriculture to help control farm pests, and provide several case studies of bushland rehabilitation in different parts of Australia. In one striking example from western Victoria, a farmer has planted over 35 000 trees and re-established wetlands on his property, with the result that there has been a dramatic increase in birdlife—from 35 species in 1956 to 148 in 1986. Pesticides have not been used for 25 years, fertiliser use is light and stocking rates are lower than average.

However, one writer (Ryan, 1990) has advocated caution in adopting tree planting programs over-enthusiastically, noting that trees can reduce yields by competing for nutrients, light and water; permanent streams may experience reduced flow regimes; some native species are not suited to nutrient-rich, 'improved' soils, and trees may limit some land-use options. Thus careful planning and monitoring is required to balance productivity needs against ecological objectives, although the Davidsons stress that the two objectives are complementary if schemes are properly designed.

PROTECTING GENETIC DIVERSITY

In response to the threat of the loss of genetic diversity, two important institutional initiatives have been launched at an international level since 1972. The International Board for Plant Genetic Resources (IBPGR) was established in 1974 under the umbrella of the Consultative Group on International Agricultural Research. The IBPGR has played a catalytic role in developing effective national and international efforts to preserve the genetic resources of major crops. Focusing largely on crops such as wheat, rice and corn, IBPGR provided technical assistance and funding to establish national and international seedbanks and to collect a large fraction of the varieties of these crops. More recently, IBPGR has shifted its priorities to crops of regional or national importance and has emphasised training and increasing human capacity (Tolba et al., 1992:207).

However, it is not often appreciated that in order to preserve seeds the crops must be grown; and this leads to a range of problems, not least of which are the incentives for the people doing the work. By what means can these efforts be maintained in perpetuity? Another is the need for duplication in a range of different environments, as insurance should a particular crop be lost for a wide range of possible reasons (including disease, fire and war), and to allow cross-breeding in order to maintain the vitality of plant varieties and species.

The second initiative took place in the early 1980s, when the Commission on Plant Genetic Resources was established in the Food and Agriculture Organisation (FAO) of the United Nations. This commission and the associated International Undertaking on Plant Genetic Resources have been instrumental in elevating concern over genetic resource ownership and access to diplomatic levels. After a decade of discussion, the principle of 'farmers' rights' is now widely accepted as a counterpoint to the concept of 'breeders' rights' (Tolba et al., 1992:207).

DIMINISHING THE ADVERSE EFFECTS OF AGRICULTURAL CHEMICALS

Probably the most important requirement to minimise the adverse effects of agricultural chemicals on the environment and human health is to reduce the quantities and frequency of their use. This requires an appreciation by the user that the economic benefits of chemical use are countered by environmental and social costs. For example, the widespread community attitude that weeds are despised plant species to be killed by a chemical treatment needs to be changed. Instead, communities should recognise that sophisticated science is required to understand the biology and ecology of weeds so that strategies may be devised for their management (Commonwealth Department of Primary Industries and Energy, 1992:8).

Pesticides

A number of countries (Sweden, Norway and Holland) and the Province of Ontario in Canada have formally set targets of halving pesticide use over the next 5 to 10 years. This is to be achieved by the strategic use of more specific and less disruptive chemicals and better methods of application, in combination with a range of other techniques. These techniques include closer monitoring (of the timing and frequency of use, and of pest densities) and various cultural (such as crop rotations), biological (use of natural predators and parasites) and biotechnological (plant breeding for pest resistance) controls. Collectively, this approach is known as Integrated Pest Management (IPM), which aims to reduce pest damage to economically acceptable levels.

Pimentel et al. (1991) have investigated the feasibility of a 50 per cent reduction of pesticide use in the United States. They concluded that food costs would rise by only 0.6 per cent with the substitution of alternative methods for existing technologies. They also noted that various studies have demonstrated that pesticide use could be reduced by one-third to one-half without adversely affecting yields. This contrasts sharply with some industry predictions that food costs would increase four- to five-fold and that food production would be halved—although these predictions unrealistically assumed a complete removal of pesticides (Krummel and Hough, 1980).

Similar concerns were expressed by the Australian Bureau of Agricultural Resources at the Senate Inquiry into Agricultural Chemicals (Senate Select Committee, 1990:119). But there is plenty of evidence that substantial reductions of pesticide use are achievable. In Australia, for example, pesticide use has been reduced to as little as one-third the level of conventional use in some sectors through the use of IPM programs. New South Wales orchardists have achieved an annual cost reduction in spider mite control from $300/ha to less than $50/ha (which is equivalent to a national saving of $12 million). The integrated management of parasites in livestock has produced potential cost savings of $5–6 million in New South Wales alone through the use of programs such as 'wormkill', 'licekill' and 'drenchkill'. A computer-aided program

(SIRATAC) used on up to one-third of the nation's cotton crop has reduced 'hard' insecticide use by an annual average of 37 per cent (maximum 72%) (Senate Select Committee, 1990).

However, IPM and alternative pest management systems are not without their difficulties. Farmers may be reluctant to adopt the methods because of their complexity, the cost of expert advisers, and the economic risks associated with conversion to the new program. In particular, Australian vegetable growers (who are also the largest unit-area users of pesticides) have resisted IPM partly for these reasons, and also because of a perceived low consumer tolerance for blemished food. In fact, IPM is as applicable to this sector as to any other. Better grower education, aided by a good backup advisory service, will be required (Senate Select Committee, 1990).

Fertilisers

The amounts of fertilisers used in agriculture can also be reduced substantially through a number of good management practices (Black, 1993). Careful timing and placement of applications, a better understanding of specific crop needs, and attention to growing environments (soil type, temperature and moisture regimes) can ensure a more economical use of fertilisers and efficient crop responses. Additional practices such as soil conservation measures (control banks and mulching) reduce nutrient loss through erosion and leaching, while increases in the amount of soil organic matter and soil biology will enhance the natural fertility of soil.

Veterinary chemicals

The routine or non-essential use of many veterinary chemicals can be avoided or reduced in a number of ways. These include: the maintenance of stock health (good diet, breeding and living conditions); stress avoidance (particularly overcrowding); reduced stocking rates; attention to quarantine measures, and avoidance of prophylactic use of antibiotics or indiscriminate use of growth promotants (NRC, 1989).

Avoidance of synthetic chemicals

The extreme alternative to an approach which relies wholly on synthetic chemicals for weed and pest control, and to maintain soil fertility, is one which does not use *any* synthetic chemicals at all. In Australia, and contrary to the opinion of most farmers and agricultural agencies, there is a growing number of organic and biodynamic farmers who farm successfully using minimal amounts of, or no, synthetic chemicals (A.J. and J.L. Conacher, 1982; J.L and A.J. Conacher, 1991).

Among the Conachers' findings were that methods designed to be 'environment friendly' from one point of view—pest or weed control, for example—may be deleterious in other ways; for example, there are adverse effects of repeated cultivations on soil properties relevant to erosion resistance, or of stubble burning on nutrient cycling. Another important finding was that even hard-core organic farmers find it very difficult to be totally purist in their approach. Organic farming (as is true for any type of farming) is not easy, and is made more difficult by the lack of the support structures enjoyed by conventional farmers. More to the point of this chapter, however, is the fact that the solution to the environmental and health problems of residues in the environment is unlikely to lie in a total ban on synthetic chemicals. It is much more realistic to aim for their controlled use, in association with other management methods.

DEALING WITH SOIL DEGRADATION

The range of problems discussed in chapters 1 and 3 included the formation of subsoil hardpans, accelerated erosion by wind and water, acidification and nutrient depletion. A number of measures have been developed to deal with these and other forms of soil degradation. They include stubble retention, reduced tillage and appropriate machinery, windbreaks, grade banks, modified rotation practices incorporating deeper-rooted legumes, liming, gypsum treatments, deep ripping, controlled grazing, and 'organic' methods.

Stubble retention—retaining crop residues in the paddock following harvesting—is designed to return nutrients to the soil. Stubble should not be burnt or removed. The incorporation of stubble in the soil also helps to prevent the formation of subsoil hardpans, by adding organic matter and improving soil structure. Surface stubbles and mulches also aid in reducing wind erosion, especially during the hot, dry summer months; and it will protect the soil from *rainwash* (accelerated erosion caused by raindrop impact and overland flow). For example, soil losses of 100 t/ha on a burnt stubble plot in tropical Queensland were reduced to negligible amounts when stubble was retained under zero tillage (Seymour, 1980).

Problems with stubble retention have included: the clogging of cultivation equipment; poor germination; increased pests and diseases, which increase the need for pesticides; and nitrogen depletion, which results in the increased use of nitrogenous fertilisers. Careful management should remove or reduce these difficulties.

Reduced tillage and the use of appropriate tillage machinery are designed to minimise both the pulverising and the compaction of the soil. Reduced tillage can be implemented without the use of herbicides, but some strategic use of the chemicals may be needed from time to time.

Windbreaks have benefits other than the direct reduction in windspeed and turbulence and hence erosivity. Loss of moisture from the soil through evaporation is also reduced, resulting in yield increases (except in the strip of land immediately adjacent to the trees), and stock use the windbreaks for much-needed shelter from sun, winds and storms—with a resultant reduction in stock losses. Commercially useful trees and shrubs can be used to construct windbreaks, thereby providing the farmer, in time, with various timber products (fencing and building materials, firewood) and, less commonly, oils, resins, stockfeed, fruits and nuts, for example. Machinery suited to large-scale planting of tree seedlings and seeds is now available to farmers. This prospect leads into the field of agroforestry, which is discussed further below.

Grade banks are standard erosion control measures which intercept overland flow, often directing it into a farm dam. They are constructed with a grader on a very slight gradient of about 1 per cent (so as not to cause erosion), with a mound of soil on the downslope side. Some form of erosion control is essential on any sloping land (which is most land), yet it is surprising and disturbing to see that many farmers do not use grade banks (or some alternative), or still do not plough on the contour. Alternatives to grade banks include contour ploughing, especially using the chisel plough which cuts deep into the soil without inverting it, or retaining uncultivated (and vegetated) strips on the contour.

Modified rotation practices and deep-rooted legumes are designed to 'rest' the soil from nutrient extraction by a particular plant season after season. Legumes return nitrogen to the soil and (if ploughed in) provide other nutrients, organic matter and 'body' to the soil. Ploughing also breaks pest and disease cycles. A legume phase also reduces water runoff and soil erosion. In parts of Australia where more than one crop per year is possible, as in northern New South Wales and southern Queensland, where rainfall is fairly evenly distributed between summer and winter, ploughing in legumes is an attractive proposition. Significant yield increases in the following crop have been measured. But in other areas, such as the 'Mediterranean' climates of southwestern and southern Australia, where only one crop per year can be grown, this practice may be an expensive luxury. On the other hand, failure to return nutrients and organic matter to the soil can result in progressively declining yields; thus the absence of resting or legume phases in the crop rotation pattern may be a false economy. As always, farmers have to juggle considerations of economics, soil quality and climate, usually in the absence of reliable data.

Liming is designed to raise the pH of a soil—that is, to render it less acid. As indicated in chapter 3, acid soils may lead to a wide range of problems. However, care needs to be taken in applying lime, as over-use can cause deficiencies of copper and iron. Application over large areas can be costly; also, treatment of surface soil horizons with lime does not deal with subsoil acidity.

Gypsum is commonly applied where there is a problem of soil sodicity—excess sodium—in heavy soils. Sodium ions disperse clay particles in the soil, resulting in a loss of soil structure and hence reduced movement of water and, especially, air through the soil. Where soils have high contents of clay-sized particles, this may also lead to 'pugging' in areas trampled by stock. Used in the correct quantities, the calcium ions in the gypsum will displace the sodium ions, resulting in the re-aggregation of clay-sized particles and the return of structure to the soil. However, excessive gypsum treatments will have the opposite effect to that intended, again causing the dispersal of the clay particles.

Deep ripping has been used to break up hardpans, aerate the soil, improve soil drainage, enable root penetration, mix acid subsoils with more neutral surface horizons, and reduce high salt concentrations at the soil surface. However, this is a very 'physical' response to a range of processes responsible for the various problems, and it would need to be repeated from time to time in the absence of methods directed at the causal processes. Not surprisingly, yield responses to this treatment have been variable.

Controlled grazing is a particularly appropriate means of dealing with degraded rangelands. It may be combined with revegetation, relocation of watering points or of stock from 'soft spots' which are vulnerable to unusually high concentrations of stock, and control of woody plant species and feral animals. Appropriately located and maintained fencing is crucial.

'*Organic*' methods have been referred to above in the context of reducing the use of synthetic pesticides. However, as pointed out by A.J. and J.L. Conacher (1982), it is important to recognise that there are both constructive and passive aspects to organic farming. The passive one is the minimisation or avoidance of synthetic chemicals. The constructive aspect involves an attempt to simulate nature and to use natural or non-synthetic materials, with a focus on building up and maintaining soil fertility and ensuring long-term sustainability. These aspects are reflected in the widely adopted United States Department of Agriculture (1980:9) definition of organic farming:

Organic farming is a production system which avoids or largely excludes the use of synthetically-compounded fertilisers, pesticides, growth regulators and livestock food additives. To the maximum extent feasible, organic farming systems rely on crop rotation, crop residues, animal manures, green manures, off-farm organic wastes, mechanical cultivation, mineral-bearing rocks and aspects of biological control to maintain soil productivity and tilth, to supply nutrients and to control insects, weeds and other pests.

This is an ecological approach to farming (see further, Altieri, 1983; NRC, 1989; Edwards et al., 1990; Lampkin, 1990), which has parallels in some of the more recent approaches to conventional farming discussed below.

Despite the general lack of institutional support (with some exceptions, notably in Victoria) for organic farming in Australia, in recent years organic farmers have responded to a rapidly growing demand for their produce, and have established a national set of standards for organically grown food (J.L. and A.J. Conacher, 1991). In 1989/90, the domestic value of the organic food market was valued at between $35 million and $45 million, with further opportunities available for exporting 'clean' food (Hassall and Associates, 1990). In view of the odds, these figures represent a fair measure of 'success' for organic growers (even though they represent only about 1% of all farmers). The value of such specialised or 'niche' markets was acknowledged by the Senate Committee inquiring into agricultural chemicals. The committee recommended that funds be provided for further research into the development of organic (and sustainable) agriculture and the assessment of export markets (Senate Select Committee, 1990).

Increasingly, the boundaries of organic or alternative agriculture are being blurred and extended by the growing use of other terms and approaches discussed further below. They include agroforestry, whole farm planning, integrated or total catchment management, landcare, and sustainable agriculture. Most of these approaches explicitly address many land degradation problems while embracing the good management practices which underpin organic methods.

OTHER APPROACHES TO BETTER MANAGEMENT

Agroforestry

From the discussion so far it will be apparent that revegetation of land is an important component of a number of remedial measures. In order to re-establish wildlife habitats, it is important to revegetate with indigenous species, and to ensure the re-establishment of total plant communities (including ground covers and shrubs), not just trees. But for other purposes, it is not essential that native species be used (although often they are the most appropriate); rather, it becomes a matter of ensuring that farmers can at least maintain their economic returns from losing crops (or pastures) to trees.

Thus it follows that trees (and shrubs and ground covers) need to be integrated with other farming practices (termed 'agroforestry') *and*, preferably, be able to provide direct commercial benefits themselves (Reid and Wilson, 1985). Agroforestry goes far beyond merely combining pine trees with some limited grazing, which the concept seems to be restricted to in some quarters. Table 7.2, for example, lists a number of reasons for practising agroforestry under semi-arid conditions, and also includes some possible species and commercial uses. A useful discussion on agroforestry's role in helping address the problem of land degradation in Australia has been provided by Carne (1993).

Table 7.2. **Benefits of agroforestry in semi-arid conditions.**

Controlling salinity by combining catchment, midslope and valley plantings with crop and water management strategies; eucalypt spp., casuarina, tamarisk

Reducing the impact of droughts by growing deep-rooted perennial fodder species; tamarisk, tagasaste, carob, honey locust, mulberry: wide range grazed by stock;

Producing shrub and tree-crop products:

commercial timber—posts, firewood, sandalwood, construction timber, wood pulp

food—nuts (walnuts, almonds, pecans); honey; fruits (avocado, fig, apricot, mulberry)

fodder—tagasaste, carob, acacias

oils—olive, candlenut, eucalypts

flowers and seeds—banksia, dryandra, various wildflower spp.

medicinal—aromatic oils (eucalypt, tea tree), alkaloids (many spp.)

other commercial products—tannins, dyes, gums, resins, fibre (acacias, eucalypts)

Reducing wind erosion hazards by planting permeable windbreaks and widely spaced trees, either covering problem sites or directly associated with crops and stock within paddocks

Protecting stock by planting trees and shrubs for shade, stress reduction, protection during lambing/calving, provision of fodder

Conserving flora and fauna by protecting habitat, preserving gene pools, and providing sites for natural predators

Protecting water catchments

Controlling water erosion—preventing erosion, siltation and flooding by planting on or upslope of affected areas

Providing cover on salt-affected land, which in turn reduces erosion and provides some fodder

Improving aesthetics, recreation and tourism, thereby providing wider community benefits (also true of many other benefits)

Increasing nutrient cycling by adding litter to the soil surface and deep root penetration; adding nitrogen by planting leguminous species, including tamarisk, tagasaste, carob and honey locust (which also provide fodder)

Adding value to property

Whole-farm and integrated catchment planning

It has become increasingly apparent over the years that attempts to solve specific land degradation problems in isolation from either other problems, or the entire range of farm management practices, are fraught with danger. Single solutions to single problems may well lead to further, unanticipated difficulties. Similarly, solutions worked out for a particular paddock are generally inappropriate for others. The entire farm has to be seen as an integrated, operating system. Often this has meant re-aligning paddock boundaries from their original grid patterns so as to take into account the natural features of the landscape (varying soil types, drainage lines, rocky outcrops and so on); or to avoid directing stock through inappropriately located gates on to particularly erodible soils, for example. Land capability evaluations provide a framework for such management procedures (Roberts, 1992: see also Kubicki et al., 1993).

The idea of integrated farm management has been extended to total catchments, for essentially the same reasons. Requiring the co-operation of a number of farmers, the concept is termed Integrated Catchment Management (ICM) or Total Catchment Management (TCM). It is particularly appropriate for degradation problems linked by running water, such as erosion, sedimentation, eutrophication or secondary soil and water salinisation.

Two examples of government initiatives come from New South Wales and Western Australia. *The Catchment Management Act* of New South Wales draws together ten government departments and other groups to identify issues and implement TCM policy, working from the community level through to state committees. In Western Australia, the 1992 *Decade of Landcare Plan* (Butcher, 1992:v) outlines the government's policy for ICM as one which 'promotes sustainable land and water practices [ensuring that] environmental values are maintained or improved'. The proposed structure will be based on a co-ordinating group of community representatives accounting for all interests in the catchment, and a group of government advisers who are to offer financial and technical support. An Office of Catchment Management promotes and co-ordinates ICM in the state.

One of the best-known examples of catchment management in Australia is centred on the Murray/Darling Basin, an area which extends over several states (Queensland, New South Wales, Victoria and South Australia) and the Australian Capital Territory, and which encompasses over 70 per cent of Australia's irrigated area. A detailed study of the basin identified a wide range of inappropriate practices and land degradation problems. In response, the Murray/Darling Basin Natural

Resources Management Strategy was set up in 1989 to co-ordinate government and community action and implement management programs through the promotion of better land-use practices and the treatment/prevention of land degradation processes.

But there has been a tendency for integrated (or total) catchment management to become 'the' solution to all problems; and it needs to be recognised that not all problems are necessarily amenable to a catchment approach. Examples include wind erosion and various soil problems, such as acidity, water repellency or loss of structure.

There are several other concerns with TCM, such as the need to obtain the willing co-operation of all landowners in the catchment, the fact that catchment boundaries don't coincide with the social boundaries of rural communities, the means of achieving an effective exchange of information, and the dominance of government agencies—which don't always wish to share their expertise (Martin and Lockie, 1993). On the other hand, it has been observed that groups formed on a catchment basis often have closer cohesion and more substantial bases for community action than other groups, because of the similarity of group members' problems (Roberts, 1992).

Soil Conservation Districts and Landcare

Closely related to the integrated catchment management concept was the development of Soil Conservation Districts, a concept which appears to have originated in New South Wales (Knowles, 1979). Then termed the 'soil conservation project concept', the approach involved the installation of soil conservation structural works together with the introduction of necessary land management practices over a whole catchment or sub-catchment, in co-operation with farmers and local authorities. In 1979 Victoria had nine such projects in operation, New South Wales 12 and Queensland 70: the total area covered was 824 000 ha, involving 3236 landholders. New South Wales had the largest area covered by these projects, with 380 000 ha.

By 1992, some 80 per cent of Western Australian farmers were involved in about 140 Soil Conservation Districts (Butcher, 1992). The districts are intended to be farmer based, but with the active involvement and encouragement of personnel from the Department of Agriculture, as well as local governments. Each district tends to focus on the particular degradation problems distinctive to that area. Boundaries may focus on shire, local community or catchment areas. Where there are over-arching problems such as salinity or flooding, the focus is on the catchment or sub-catchments.

The Soil Conservation Districts seem to have become merged (or confused?) with the Landcare concept. Landcare was a set of initiatives proposed by the National Farmers' Federation and the Australian Conservation Foundation, and adopted by the federal government (see below). This, too, involves farmers grouping together to tackle common problems in association with agency personnel and local governments, but the areas are often a lot smaller than the districts. Thus a Soil Conservation District could encompass a number of Landcare groups.

NFF/ACF National Land Management Program

In 1989 the National Farmers' Federation (NFF) and the Australian Conservation Foundation (ACF) submitted a joint proposal to the federal government for a national land management program. They proposed that agricultural and pastoral lands be used within their capabilities by the year 2000 and that there be sustainable use of those lands after that date (Toyne and Farley, 1989). The proposal sought a major injection of funds and the support of rural-based land management groups, and several other initiatives, as a means of providing momentum for action on land degradation. In essence, the proposal was adopted in the *Prime Minister's Statement on the Environment* of July 1989 and the announcement of the Decade of Landcare (Hawke, 1989).

The *Decade of Landcare* was to provide over $320 million to the year 2000 for Landcare, tree planting, vegetation protection, a review of rehabilitation policies (such as taxation and drought relief), and restructuring of the National Soil Conservation Program.

A new corporation, the National Resources Development Research Corporation, was established to examine soil, water and forestry issues. Several tree programs received major funding: the One Billion Trees Program (revegetation, seeding and community involvement); Save the Bush (protection of natural remnant vegetation outside national parks), and the National Reafforestation Program (to encourage the establishment of hardwood plantations for forestry and rehabilitation purposes). Additional funds were made available to the Murray/Darling Basin Commission to assist the development and implementation of its management strategy.

Attention was also given to upgrading agricultural chemicals legislation and the monitoring of food for residues.

By 1993 there were over 1400 Landcare groups around Australia (figure 7.2), of which more than half had been established for less than two years. According to a 1992/93 national survey, membership of the Landcare groups was estimated to be around 28 per cent for broadacre farms and 19 per cent in dairying, also ranging between the states/territories from a high of 47 per cent in the Northern Territory to 12 per cent in Tasmania. In general, in comparison with all farmers, members of these groups tend to be younger, more innovative and willing to take risks, more likely to have good farm plans and to be aware of degradation problems on their properties, and to manage larger properties (Campbell, 1992; Mues et al., 1994). Given the immense social and economic pressures on rural areas, Landcare has become an important social focus. The 1992/93 Landcare survey noted that farmers put a high value on demonstrations and field days (Mues et al., 1994).

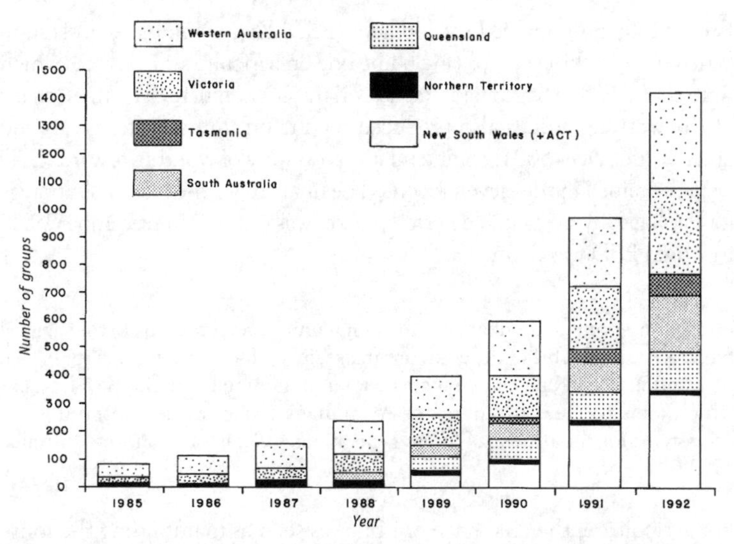

Figure 7.2 The growth of Landcare groups in Australia, by state and territory, from 1985 to 1992.

Source: Campbell 1992: fig. 5.1.

While these initiatives have been very encouraging, there are some concerns. Chief among them is the possibility that the problem the group was formed to deal with may not be solved. Since it has taken a long time for the idea of spatial (as well as group) co-operation to take hold, and in some areas for suspicions to disperse, failure would have ramifications which would last a long time. Currently, it appears from discussions with a number of people that, as one might expect, some groups are performing very well, many are making some progress, and another minority are doing very little or have lost momentum. Much depends on how well a problem is targeted, the quality of leadership and training, the availability of funds and the efficient transfer of information. In addition, the prospect needs to be faced that solutions of some forms of land degradation may not be known. This prospect is aggravated by the fact that rarely does one form of degradation exist on its own.

Ecologically sustainable land management

The concept of 'sustainable development' evolved from the Brundtland Report, Our Common Future (World Commission on Environment and Development, 1987). In 1990, the Australian government released an

Ecologically Sustainable Development (ESD) Discussion Paper and established nine working groups (including one on agriculture) in sectors which were considered to have major impacts on the environment. The three chairs of the working groups also produced reports on intersectoral issues and greenhouse. Over 500 recommendations on ways of working towards ESD were contained in the eleven reports. The final report, the National Strategy for Ecologically Sustainable Development, was released in December 1992, in which ESD was defined as:

> using, conserving and enhancing the community's resources so that ecological processes, on which life depends, are maintained, and the total quality of life, now and in the future, can be increased. Put more simply, ESD is development which aims to meet the needs of Australians today, while conserving our ecosystems for the benefit of future generations. (Commonwealth of Australia, 1992:6)

For agriculture, the challenge of ESD was seen as maintaining the long-term productivity of agriculture and its contribution to Australia's economic well-being, while protecting the biological and physical resource base and improving human health and safety.

Support is given in the *National Strategy* to a wide range of actions, including: Landcare; ICM and TCM; farm plans and community groups; improved access to funds for groups seeking assistance under resource management programs; revegetation and remnant protection programs; the use of taxation incentives; IPM, and the National Strategies on Weeds and the Management of Vertebrate Pests; the control of introduced animals and plants; more efficient use and pricing of water; implementation of recommendations from the Senate's chemicals inquiry; improvements in technical advice and training; the integration of institutional arrangements; liberalisation of world trade; coping with greenhouse; and monitoring and review of the ESD program (Commonwealth of Australia, 1992).

These and many other recommendations refer not only to agriculture and forestry but to a number of intersectoral issues, including biological diversity, nature conservation, native vegetation, environmental protection, land use planning and decision making, natural resource and environment information and environmental impact assessment. The *National Strategy* is accompanied by a *Compendium of ESD Recommendations*. This is an important document which puts flesh on the *National Strategy* framework. It lists all the working groups' and chairs' recommendations; relates the governments' responses, with examples of relevant existing and proposed policies, programs and actions, to the recommendations; identifies bodies with primary responsibility for implementation, and sets out some time-frames for implementation.

WHO PAYS?

This is an obvious question but with no obvious answer. Although throwing money at a problem doesn't necessarily solve it, it is clear that a lot of money will need to be spent on dealing with Australia's land degradation problems. Where should the money come from?

If one accepts the 'polluter pays' principle, then the farmers should pay. They have caused or are causing the damage, directly or indirectly, wittingly or unwittingly. But, as chapter 6 has shown, many farmers are in no position to incur additional expenditures.

If one accepts the 'user pays' principle, then all those who stand to benefit from the repair of land degradation should contribute to the costs of doing so. Arguably, this applies to all Australians—through an improved environment for future generations, improved yields and therefore an improved balance of payments (from exported produce) or improved security of supply (for locally consumed produce). Therefore, 'the government' (= the taxpayer) should pay. But as we all know, Australia is living beyond its means, with huge budget deficits: like the farmers, the country cannot afford the price. Nevertheless, it has been demonstrated that the community is willing to bear some sort of surcharge to fund conservation programs (Dragovich, 1990, 1991).

Beneficiaries also include overseas consumers, who normally would pay through the price of the product. But Australia does not compete on equal terms with other countries for overseas markets. Other countries—notably western Europe and the United States in recent years—subsidise their produce on overseas markets. Australia cannot load the costs of repairing land degradation on to the prices of its export produce: those prices are already too high for many markets. There have been suggestions that the costs of land degradation could be 'internalised' by taxes on farm chemicals, through quotas, or by making alternative farming methods available at lower cost.

Where does this leave us? Realistically, it means that land degradation will not be solved in the short term through major, expensive projects. Rather, it will take a long time, involving patient research by government agencies, CSIRO and universities, policy development and implementation, and farmer education and incentives, to improve gradually the way we manage our land until our farming practices become truly sustainable.

CONCLUSION

'Ecologically sustainable' agriculture goes a long way towards addressing the problems of conventional practices and the legacies of the past. But it is not a magic solution to land degradation.

Management concepts (contained in various farm, district or catchment planning methods and computerised packages) can be complex and difficult for many farmers to implement without good community and technical support. Farmers and administrators deal with burgeoning bureaucratic structures—in fact, many of the new institutional arrangements seem to be merely replacing one agency with another. According to Farley (the past president of the National Farmers' Federation), there are over 100 Commonwealth government agencies dealing with different aspects of Landcare and at least ten ministerial committees. In 1986, there were 38 pieces of Commonwealth legislation affecting land use (and another 38 in New South Wales alone). On top of that, there are over 800 local governments with substantial powers, as well as the activities of various research institutions and community groups (Farley, 1990, 1991). Not surprisingly, there is an immense duplication of effort and a disproportionate number of administrators in relation to the work on the ground.

There are difficulties relating to funding. The increased availability of funds through the Land and Water Resources Council (which incorporates the older National Soil Conservation Program) has been seen by some as a 'honeypot', and in some cases has been wasted on ill-conceived projects. In other cases, it has been seen as an opportunity for government agencies and researchers to fund themselves at the farmers' expense—although present Landcare policies try to ensure that funding goes to the grassroots, community level. There is a need for the various projects to be subject to a review process, unfortunately involving more expense and unproductive personnel, and to ensure that the information goes out to those who will benefit most from it.

Not all solutions to land degradation problems are simple economic or technical ones. Social and environmental (or 'external') costs need to be accounted for in rigorous terms.

Data are still a big problem. Inaccurate information based on shaky evidence continues to be peddled in the literature: many writers repeat wrong information from secondary sources and fail to check the original material. The emphasis of research in agricultural education at tertiary level, especially in universities, has remained very much on production science rather than on resource or environmental issues. There is a great need for properly trained personnel to assist both bureaucrats and farmers.

Farmer responses to various land management programs have been patchy, reflecting in part the need to secure attitudinal changes among farmers through better education processes and more effective transfer of technical information. Even so, some farmers complain of information overload, a problem which also needs attention.

Farmers continue to be pressured by economic constraints and, as a result, defer crucial conservation measures, thereby exacerbating land degradation problems.

For all these concerns, there is no question that there have been enormous advances over the last decade in the recognition and treatment of land degradation in Australia. This was given particular impetus at the end of the 1980s with the National Soil Conservation Program, the *Prime Minister's Statement on the Environment*, and the *National Strategy for Ecologically Sustainable Development*. Rapid growth in Landcare and catchment groups resulted, and tree planting programs were accelerated. The fear is that much of this enthusiasm will wane as a result of insufficient skilled technical and political support and leadership, too many institutional barriers, and preoccupation with other economic priorities.

■

References

ABARE (1991) *Farm surveys report: financial performance of Australian farms.* Australian Bureau of Agricultural and Resource Economics, Australian Government Publishing Service, Canberra.

ABARE (1992) *Commodity statistical bulletin.* November 1992, Australian Bureau of Agricultural and Resource Economics, Australian Government Publishing Service, Canberra.

Abbott, I. and Black, R. (1987) 'Diversity of terrestrial invertebrates on islets: interrelationships among floristics, vegetation, microhabitats and sampling effort'. In A.J. Conacher (ed.), *Readings in Australian Geography: Proceedings of the 21st Institute of Australian Geographers' Conference* (pp. 476–81). Institute of Australian Geographers (WA Branch) and Department of Geography, University of Western Australia, Nedlands.

Abbott, I., Parker, C.A. and Sills, I.D. (1979). 'Changes in abundance of large soil animals and physical properties of soils following cultivation'. *Australian Journal of Soil Research*, 17: 343–53.

Abrahams, A.D. (ed.) (1986) *Hillslope processes.* Allen & Unwin, Boston.

ABS (1986) *Agricultural and selected inputs 1985–6.* Australian Bureau of Statistics, Catalogue 7411.0, Commonwealth of Australia, Canberra.

ABS (1991) *Australian Yearbook 1991.* Commonwealth of Australia, Canberra.

ABS (1992) *Australia's environment: issues and facts.* Australian Bureau of Statistics, Catalogue No. 4140.0, Commonwealth of Australia, Canberra.

Adamson, D., Selkirk, P.M. and Mitchell, P. (1983) 'The role of fire and Lyre birds in the sandstone landscape of the Sydney Basin'. In R.W. Young and G.C. Nanson (eds), *Aspects of Australian sandstone landscapes* (pp. 81–93). Australian and New Zealand Geomorphology Group Special Publication No. 1, University of Wollongong.

Allison, G.B., Cook, P.G., Barnett, S.R. and others (1990) 'Land clearance and river salinisation in the western Murray Basin'. *Journal of Hydrology*, 119: 1–20.

Altieri, M.A. (1983) *Agroecology: the scientific basis of alternative agriculture*. Division of Biological Control, University of California, Berkeley.

Anon. (1983) 'Overview'. *Quarterly Review of the Rural Economy*, Bureau of Agricultural Economics, 5(1): 4–7.

Anon. (1984) 'Policies that accelerate land degradation'. *Search*, 15: 130.

Anon. (1985) 'Agricultural courses in Australia'. In J. Cribb (ed.), *National Farmers' Federation Australian agricultural yearbook* (pp. 509–14). Strand Publishing, Brisbane.

Anon. (1992) 'Land and trees'. *Our Planet* (United Nations Environment Program), 4(2): 6.

Anon. (1993) 'Surprising results from farmer surveys, Landcare Australia news'. *Australian Journal of Soil and Water Conservation*, 6(3): 57.

ANOP (1992) *The environment and the ESD process: an attitude research analysis*, Volume 1 of Report on 1991 National Research Program prepared for the Department of ASET and the ESD Secretariat. ANOP Research Services Pty Ltd, North Sydney.

Benson, J. (1990) 'Australia's threatened plants'. In M. Kennedy (ed.), *Australia's endangered species: the extinction dilemma* (pp. 145–51; see also Appendix IV, pp. 173–85 ('The Threatened Plants') for a comprehensive list of species). Simon and Schuster, Brookvale, New South Wales.

Black, C.A. (1993) *Soil fertility, evaluation and control*. Lewis Publishers, Boca Baton, Florida.

Blaikie, P. and Brookfield, H. (1987) 'Questions in history in the Mediterranean and Western Europe'. In P. Blaikie and H. Brookfield (eds), Land Degradation and Society (pp. 122–42). Methuen, London.

Bligh, B. (1980) *Cherish the earth: the story of gardening in Australia*. David Ell Press, Sydney.

Bolton, G. (1981) *Spoils and spoilers: Australians make their environment 1788–1980*. Allen & Unwin, Sydney.

Boon, S. and Dodson, J.R. (1992) 'Environmental response to land use at Lake Curlip, East Gippsland, Victoria'. *Australian Geographical Studies*, 30(2): 206–21.

Bowie, I.J.S. and Smailes, P.J. (1988) 'The country town'. In R.L. Heathcote (ed.), *The Australian experience* (pp. 233–56). Longman Cheshire, Melbourne.

Bradsen, J.R. (1988) *Soil conservation legislation in Australia*, Report for the National Soil Conservation Programme. University of Adelaide.

Breckwoldt, R. (1983) *Wildlife in the home paddock: nature conservation for Australian farmers*. Angus and Robertson, Sydney.

Briggs, J.D. and Leigh, J.H. (1988) *Rare and threatened Australian plants*, Special Publication 14. Australian National Parks and Wildlife Service, Canberra.

Brown, A.W. (1978) *Ecology of pesticides*. Wiley, New York.

Brown, L.R. (1981) *Building a sustainable society*. Worldwatch Institute. W.W. Norton and Co., New York.

Brown, L.R. (1991) 'The new world order'. In L.R. Brown et al. (eds), *State of the world 1991*, Worldwatch Institute. Allen & Unwin, Sydney.

Bryan, R.B. (ed.) (1987) *Rill erosion: processes and significance*, Catena Supplement 8. Catena Verlag, Cremlingen.

Bryant, L. (1992) 'Social aspects of the farm financial crisis'. In Lawrence et al. pp. 157–72.

Bunney, S (1990) 'Prehistoric farming caused "devastating" soil erosion'. *New Scientist*, 125: 17.

Bureau of Agricultural Economics (1985) 'Statistics of Australian agriculture'. In J. Cribb (ed.), *National Farmers' Federation Australian agricultural yearbook* (pp. 153–67). Strand Publishing, Brisbane.

Bureau of Rural Resources (1989) 'Submission to the Senate Select Committee on Agricultural and Veterinary Chemicals'. *Hansard*, 19 July: 2787–857.

Burnley, I.H. (1988) 'Population turnaround and the peopling of the countryside? Migration from Sydney to country districts of New South Wales'. *Australian Geographer* 19: 268–83.

Burt, T.P., Heathwaite, A.L. and Trudgill, S.T. (eds) (1993) *Nitrate: processes, patterns and management*. John Wiley and Sons, Chichester, England.

Butcher, E. (chair) (1992) *Decade of Landcare Plan, Western Australia. An action program for sustainable use of agricultural and pastoral lands*. Soil and Land Conservation Council of Western Australia, Perth.

Butzer, K.W. (1974) 'Accelerated soil erosion: a problem of man:land relationships'. In I.R. Manners and M.W.M. Mikesell (eds), *Perspectives on Environment*. Association of American Geographers Publication 13, Washington DC.

California Department of Food and Agriculture (1984) *Summary of reports from physicians of illnesses that were possibly related to pesticide exposure during the period of Jan. 1–Dec. 31, 1983, in California*. California Department of Food and Agriculture, Sacramento.

Cameron, J.M.R. (1977) 'Coming to terms: the development of agriculture in pre-convict Western Australia'. *Geowest* No. 11, Department of Geography, University of Western Australia.

Campbell, A. (1992) *Landcare in Australia: taking the long view in tough times*, 3rd Annual Report of the National Landcare Facilitator. National Soil Conservation Program, Queanbeyan, New South Wales.

Carder, D.J, and Humphry, M.G. (1983) 'The costs of land degradation'. *Western Australian Journal of Agriculture*, 24: 50–3.

Carne, R.J. (1993) 'Agroforestry land use: the concept and practice'. *Australian Geographical Studies*, 31(1): 79–90.

Cary, J. and Barr, N. (1992) 'The semantics of forest cover: how green was Australia?' In Lawrence et al., pp. 60–76.

Chatterton, L. and Chatterton, B. (1982) 'The politics of pastoralism'. *Habitat*, 10: 12–14.

Chepil, W.S. and Woodruff, N.P. (1963) 'The physics of wind erosion and its control'. *Advances in Agronomy*, 15: 211–302.

Chisholm, A. and Dumsday, R. (eds) (1987) *Land degradation: problems and policies*. Cambridge University Press, Cambridge.

Clarke, A.L. (1986) 'The impact of agricultural practices on Australian soils: cultivation'. In J.S. Russell and R.F. Isbell (eds), *Australian soils: the human impact* (pp. 273–303). University of Queensland Press, St Lucia.

Cocks, D. (1992) *Use with care: managing Australia's resources in the twenty first century*. New South Wales University Press, Kensington.

Commonwealth Department of Primary Industries and Energy (n.d. (November 1992)) *Towards a national weeds strategy*. Canberra.

Commonwealth of Australia (1992) *National strategy for ecologically sustainable development*. Australian Government Publishing Service, Canberra.

Conacher, A.J. (1978) 'Resources and environmental management: some fundamental concepts and definitions'. *Search*, 9: 437–41.

Conacher, A.J. (1982) 'Dryland agriculture and secondary salinity'. In W. Hanley and M. Cooper (eds), *Man and the Australian environment* (pp. 113–25). McGraw-Hill, Sydney.

Conacher, A.J. (1986) 'Environmental conservation'. In D.N. Jeans (ed.), *Australia—a geography*, vol. 1: *The natural environment*, 2nd edn (pp. 315–39). Sydney University Press, Sydney.

Conacher, A.J., Combes, P.L., Smith, P.A. and McLellan, R.C. (1983) 'Evaluation of throughflow interceptors for controlling secondary soil and water salinity in dryland agricultural areas of southwestern Australia: I. Questionnaire surveys'. *Applied Geography*, 3: 29–44.

Conacher, A.J. and Conacher, J.L. (1982) *Organic farming in Australia*, Geowest 18. Department of Geography, University of Western Australia. 64 pp.

Conacher, A.J. and Conacher, J.L. (1988) 'The exploitation of the soils'. In R.L. Heathcote (ed.), *The Australian experience* (pp. 127–38). Longman Cheshire, Melbourne.

Conacher, J.L. and Conacher, A.J. (1986) *Herbicides in agriculture: minimum tillage, science and society*, Geowest 22, Occasional Papers. Department of Geography, University of Western Australia. 169 pp.

Conacher, J.L. and Conacher, A.J. (1991) 'An update on organic farming and the development of the organic industry in Australia'. *Biological Agriculture and Horticulture*, 8: 1–16.

Costin, A. (1991) 'Land use and water quality. The importance of soil cover'. *Australian Journal of Soil and Water Conservation*, 4(3): 12–17.

Costin, A.B. and Frith, H.J. (eds) (1974) *Conservation*, rev. edn. Penguin, Ringwood.

Courtenay, P.P. (1988) 'The tropical experience'. In R.L. Heathcote (ed.), *The Australian experience* (pp. 85–95). Longman Cheshire, Melbourne.

Cribb, A.B. and Cribb, J.W. (1981a) *Wild medicine in Australia*. Collins, Sydney.

Cribb, A.B. and Cribb, J.W. (1981b) *Useful wild plants in Australia*. Collins, Sydney.

Cribb, J. (1982) *The forgotten country*. Australasian Farm Publications, Melbourne.

Cribb, J. (1989) 'Agriculture in the Australian economy'. In J. Cribb (ed.), *Australian agriculture: the complete reference on rural industry*, vol. 2 (pp. 11–48). Morescope, Camberwell, Victoria.

Cribb, J. (ed.) (1991) *Australian agriculture: the complete reference on rural industry*, 3rd edn. Morescope, Camberwell, Victoria.

Cribb, J. (1992) 'Sewage main culprit in algal blooms'. *The Australian*, 30 January.

CSIRO (1975) 'Keeping a watch on cadmium'. *Ecos*, 5: 29–31.

CSIRO (1980) 'Troubles aplenty in our cotton fields'. *Ecos*, 23: 10–12.

CSIRO (1990) *Australia's environment; its natural resources. An outlook*. CSIRO, Canberra.

Cullen, P. (1991) 'Land use and declining water quality'. *Australian Journal of Soil and Water Conservation*, 4(3): 4–8.

Cullen, T., Dunn, P. and Lawrence, G. (eds) (1989) *Rural health and welfare in Australia: Proceedings of the Conference on the Crisis in Rural Health and Welfare*. Centre for Rural Welfare Research, Charles Sturt University, Riverina.

Davidson, B.R., (1967) *Australia—wet or dry?* Melbourne University Press, Melbourne.

Davidson, B.R. and Graham-Taylor, S. (1982) *Lessons from the Ord*. Centre for Independent Studies, St Leonards, New South Wales.

Davidson, R. and Davidson, S. (1992) *Bushland on farms: do you have a choice?* Australian Government Publishing Service, Canberra.

Deevey, E.S., Rice, D.S., Rice, P.M. and others (1979) 'Mayan urbanism: impact on a tropical karst landscape'. *Science*, Oct. 19: 298–306.

De Garis, B. (1979) 'Settling on the sand: the colonisation of Western Australia'. In B. De Garis (ed.), *European impact on the West Australian environment 1829–1979* (pp. 1–15). Octagon Lectures, University of Western Australia, Perth.

Department of Arts, Heritage and Environment (1986a) *State of the environment in Australia*. Australian Government Publishing Service, Canberra.

Department of Arts, Heritage and Environment (1986b) *State of the environment in Australia source book*. Australian Government Publishing Service, Canberra.

Department of Environment, Housing and Community Development (1978) *A basis for soil conservation policy in Australia: Commonwealth and State Government collaborative soil conservation study 1975–77. Report 1*. Australian Government Publishing Service, Canberra.

De Ploey, J. (ed.) (1983) *Rainfall simulation, runoff and soil erosion*, Catena Supplement 4. Catena Verlag, Cremlingen.

Diamond, J.M. (1975) 'The island dilemma: lessons of modern biogeographic studies for the design of nature reserves'. *Biological Conservation*, 7: 129–45.

Division of National Mapping (1982) *Atlas of Australian resources*. Third Series vol. 3: Agriculture. Commonwealth Government Printer, Canberra.

Dosman, J.A. and Cockroft, D.W. (eds) (1989) *Principles of health and safety in agriculture*. CRC Press, Boca Baton, Florida.

Douglas, I. (1992) 'Sources and destinations of sediment in the Huang He and Changjiang basins of China'. *Annales de Geographie*, 567: 541–52.

DPI (1980) *A manual of safe practice in the handling and use of pesticides*. Document PB377, Pesticides Section, Department of Primary Industry, Australian Government Publishing Service, Canberra.

DPIE/ASCC (1988) *Report of Working Party on effects of drought assistance measures and policies on land degradation*. Department of Primary Industry and Energy/Australian Soil Conservation Council, Australian Government Publishing Service, Canberra.

Dragovich, D. (1990) 'Does soil erosion matter to people in metropolitan Sydney?' *Australian Journal of Soil and Water Conservation*, 3(1): 29–32.

Dragovich, D. (1991) 'Who should pay for soil conservation? Community attitudes about financial responsibility for land repair'. *Australian Journal of Soil and Water Conservation*, 4(1): 47.

Dregne, H.E. (1986) 'Magnitude and spread of the desertification process'. In I.P. Gerasimov and others (eds), *Arid land development and the combat against desertification: an integrated approach* (pp. 10–16). United Nations Environment Programme, USSR Commission for UNEP, Centre for International Projects GKNT, Moscow.

Dregne, H.E., Kassas, M. and Rosanov, B. (1991) 'A new assessment of the world status of desertification'. *Desertification Control Bulletin*, 20: 6–18.

Dunin, F.X. and Mackay, S.M. (1982) 'Evaporation of Eucalypt and Coniferous forest communities'. In E.M. O'Loughlin and L.J. Bren (eds), *First National Conference on Forest Hydrology* (pp. 18–25). Melbourne.

Dunlap, R.E., Gallop, G.H. and Gallop, A.M. (1993) 'International public opinion toward the environment'. *Impact Assessment*, 11(1): 3–25.

Dunne, T. and Leopold, L.B. (1978) *Water in environmental planning*. Freeman, San Francisco.

Eckersley, R. (1989) *Regreening Australia: the environmental, economic and social benefits of reafforestation*. Occasional Paper No. 3. CSIRO, Canberra.

Edwards, C.A., Lal, R., Madden, P. and others (eds) (1990) *Sustainable agricultural systems*. Soil and Water Conservation Society, Ankeny, Iowa.

Edwards, K. (1991) 'Soil formation and erosion rates'. In P.E.V. Charman and B.W. Murphy (eds), *Soils, their properties and management. A soil conservation handbook for New South Wales* (pp. 36–47). Sydney University Press/New South Wales Soil Conservation Service, Sydney.

Edwards, K. (1993) 'Soil erosion and conservation in Australia'. In D. Pimentel (ed.), *World soil erosion and conservation* (pp. 147–69). Cambridge University Press, Cambridge.

Ellington, T., Reeves, T.G. and Peverill, K.I. (1981) 'Chlorosis and stunted growth of wheat crops in N.E. Victoria'. In *Proceedings of the Soil Management Conference* (pp. 91–109). Australian Society of Soil Science, Dookie, Victoria.

El-Swaify, S.A., Moldenhauer, W.C. and Lo, A. (eds) (1985) *Soil erosion and conservation*. Soil Conservation Society of America, Ankeny, Iowa.

Emmett, W.W. (1970) *The hydraulics of overland flow on hillslopes*. United States Geological Survey Professional Paper 662–A, Washington DC.

Emsley, J. (1992) 'Weedy gauge of ozone pollution'. *New Scientist*, 136: 15.

ESD (1991) *Greenhouse report: working group*. Ecologically Sustainable Development, Australian Government Publishing Service, Canberra.

Evans, D.G. (1991) *Acid soils in Australia: the issues for Government*. Bureau of Rural Resources, Canberra.

FAO (1991) *The state of food and agriculture*. Food and Agriculture Organisation of the United Nations, Rome.

Farley, R. (1990) 'President's message'. *Australian Journal of Soil and Water Conservation*, 3(3): 2.

Farley, R. (1991) 'President's message'. *Australian Journal of Soil and Water Conservation*, 4(2): 1.

Fergusson, J.E. (1990) *The heavy elements: chemistry, environmental impact and health effects*. Pergamon Press. London.

Foster, G.R. (1988) Modelling soil erosion and sediment yield. In R. Lal (ed.), *Soil erosion research method* (pp. 97–117). Soil and Water Conservation Society, Ankeny, Iowa.

Fry, P.W. and Goss, K.F. (1985) *Communication networks and the adoption of three farm practices*. Western Australian Department of Agriculture, Perth.

Garman, D.E.J. (1983) 'Water quality issues'. In *Water 2000: consultants report no. 7*. Australian Government Publishing Service, Canberra.

Gell, P.A. and Mueck, S.G. (1987) 'Applications of isolate biogeographic theory to the delineation and management of mallee nature reserves'. In A.J. Conacher (ed.), *Readings in Australian Geography: Proceedings of the 21st Institute of Australian Geographers' Conference* (pp. 481–92). Institute of Australian Geographers (WA Branch) and Department of Geography, University of Western Australia, Nedlands.

George, R.J. (1990) 'The 1989 saltland survey'. *Journal of Agriculture Western Australia*, 31: 159–66.

George, R.J. (1991) Interactions between Perched and Deeper Groundwater Systems in Relation to Secondary, Dryland Salinity in the Western Australian Wheatbelt: Processes and Management Options. PhD Thesis in Geography, University of Western Australia.

Gill, A.M., Groves, R.H. and Noble, I.R. (eds) (1981) *Fire and the Australian biota*. Australian Academy of Science, Canberra.

Ginnivan, D. and Lees, J. (1991) *Moving on: farm families in transition from agriculture*. Rural Development Centre, University of New England, Armidale, New South Wales.

Goodland, R. and Daly, H. (1993) *Poverty alleviation is essential for environmental sustainability*. World Bank Environmental Development Division Working Party 1993.42.

Gorddard, B. (1991) The adoption of minimum tillage in the Western Australian wheatbelt. Paper to 35th Annual Conference of the Australian Agricultural Economics Society, Armidale, New South Wales.

Government of Canada (1991) *The State of Canada's Environment*. Minister of Supply and Services Canada, Ottawa.

Graetz, D., Fisher, R. and Wilson, H. (1992) *Looking back: the changing face of the Australian continent 1972–1992*. CSIRO Office of Space Science and Applications, Canberra.

Grant, R. (ed.) (1992) *State of the environment report*. Government of Western Australia, Perth.

Green, K.D. (1983) *Water 2000: a perspective on Australia's water resources to the year 2000*, Report of the Steering Committee in conjunction with the Department of Resources and Energy. Australian Government Publishing Service, Canberra.

Guenzi, W.D., Ahlrichs, J.L., Chesters, G. et al. (eds) (1974) *Pesticides in soil and water*. Soil Science Society of America, Madison.

Gutteridge Haskins and Davey (1992) *An investigation of nutrient pollution in the Murray–Darling river system*, Report for Murray/Darling Basin Commission, Victoria.

Hallam, S.J. (1975) *Fire and hearth*. Australian Institute of Aboriginal Studies, Canberra.

Hamblin, A., and Kyneur, G. (1993) *Trends in wheat yields and soil fertility in Australia*. Department of Primary Industries and Energy, Bureau of Resource Sciences, Australian Government Publishing Service, Canberra.

Hamilton, G.J. (1970) 'The effects of sheet erosion on wheat yield quality'. *Journal of the Soil Conservation Service of New South Wales*, 26: 118–23.

Hassall and Associates (1990) *The market for Australian produced organic food*, Report by Hassall and Associates for the Australian Special Rural Research Council. Department of Primary Industries, Canberra.

Hawke, R.J.L. (1989) *Our country, our future: Prime Minister's statement on the environment*, 2nd edn. Australian Government Publishing Service, Canberra.

Heathcote, R.L. (1987) 'Pastoral Australia'. In D.N. Jeans (ed.), *Australia—a geography*, vol. 2, 2nd edn (pp. 259–300). Sydney University Press, Sydney.

Heathcote, R.L. (ed.) (1988) *The Australian experience: essays in Australian land settlement and resource management*. Longman Cheshire, Melbourne.

Hellden, U. (1991) 'Desertification—time for an assessment?' *Ambio*, 20, 372–83.

Henderson-Sellers, A. and Blong, R. (1989) *The Greenhouse Effect: living in a warmer Australia*, New South Wales University Press, Sydney.

Higgins, E. (1992) 'Crisis of our rivers'. *Australian Weekend Review*, 28–29 November: 3.

Hingston, F.J., Turton, A.G. and Dimmock, G.M. (1979) 'Nutrient distribution in karri (Eucalyptus diversicolor F. Muell.) ecosystems in southwest Western Australia'. *Forest Ecology and Management*, 2: 133–58.

Hobbs, R.J. and Saunders, D.A. (eds) (1993) *Reintegrating fragmented landscapes: towards sustainable production and nature conservation*. Springer Verlag, New York.

Hollick M. (1990) 'Land conservation policies and farmer decision-making'. *Australian Journal of Soil and Water Conservation*, 3(1): 6–13.

Holliday, V. (ed.) (1988) *Soils in Archaeology*. Smithsonian Institute Press, Phoenix.

Holmes, B. (1993) 'The perils of planting pesticides'. *New Scientist*, 28 August: 34–37.

Holmes, J.H. (1981a) 'Lands of distant promise'. In Lonsdale and Holmes (pp. 1–13).

Holmes, J.H. (1981b) 'Sparsely populated regions of Australia'. In Lonsdale and Holmes (pp. 70–104).

Holmes, J.H. (1988) 'Remote settlements' In R.L. Heathcote (ed.), *The Australian experience: essays in Australian land settlement and resource management* (pp. 68–84). Longman Cheshire, Melbourne.

House of Representatives Standing Committee on Environment, Recreation and Arts (1989) *The effectiveness of land degradation policies and programmes*. Australian Government Publishing Service, Canberra.

Houston, J.M. (1978) 'The concepts of "place" and "land" in the Judaeo–Christian tradition'. In D. Ley and M. Samuels (eds), *Humanistic Geography* (pp. 224–37). Croom Helm, London.

Hudson, N. (1981) *Soil conservation*, 2nd edn. B.T. Batsford Ltd, London.

Hughes, P.J. and Sullivan, M.E. (1986) 'Aboriginal landscapes'. In J.S. Russell and R.F. Isbell (eds), *Australian soils: the human impact* (pp. 117–33). University of Queensland Press, Brisbane.

Hussey, B.M.J. and Wallace, K.J. (1993) *Managing your bushland*. Department of Conservation and Land Management, South Perth.

IUCN/WWF (1980) *World conservation strategy*. International Union for the Conservation of Nature and the World Wildlife Fund, United Nations Environment Program.

Jacobsen, T. and Adams, R.A. (1958) 'Salt and silt in ancient Mesopotamian agriculture'. *Science*, Nov. 21: 1251–8.

Joint Committee (1979) *Ord river irrigation area review*. Australian Government Publishing Service, Canberra.

Joseph, L.E. (1990) *Gaia: the growth of an idea*. Arkana, London.

Kaloyanova, F. (1983) 'Occupational hazards of pesticides and their epidemiology'. In *Proceedings of the International Conference on Hazardous Agrochemicals*, vol. 1 (pp. 312–44). Alexandria.

Kellehear, A. (1989) 'A critique of national health policies for rural people'. In T. Cullen et al. (eds), *Rural health and welfare in Australia: Proceedings of the Conference on the Crisis in Rural Health and Welfare* (pp. 28–37). Centre for Rural Welfare Research, Charles Sturt University, Riverina, New South Wales.

Kennedy, M. (ed.) (1990) *Australia's endangered species: the extinction dilemma*. Simon and Schuster, Brookvale, New South Wales.

Kirby, M.G. and Blyth, M.J. (1987) 'Economic aspects of land degradation in Australia'. *Australian Journal of Agricultural Economics*, 31(2): 154–74.

Kirkby, M.J. and Chorley, R.J. (1967) 'Throughflow, overland flow and erosion'. *International Association of Scientific Hydrology Bulletin*, 12, 5–21.

Kirkpatrick, J.B. (1991) 'The geography and politics of species endangerment'. *Australian Geographical Studies*, 29(2): 246–54.

Knopke, P. and Harris, J. (1991) 'Changes in input use on Australian farms'. *Agriculture and Resources Quarterly*, 3(2): 230–40.

Knowles, G.H. (1979) 'Erosion assessment and control techniques in Australia'. *Journal of the Soil Conservation Service of New South Wales*, 35: 70–81.

Kondinin Group (1993) 'Personal protection equipment: cover-up tactics for farmers'. *Farming Ahead with the Kondinin Group*, 19: 27–30.

Krummel, J. and Hough, J. (1980) 'Pesticides and controversies: benefits versus costs'. In D. Pimentel and J. Perkins (eds), *Pest control: cultural and environmental aspects* (pp. 159–79). American Association for the Advancement of Science Selected Symposium 43. Westview Press, Colorado.

Kubicki, A., Denby, C., Stevens, M. et al. (1993) 'Determining the long term costs and benefits of alternative farm plans. In Hobbs and Saunders (pp. 245–78).

Lampkin, N. (1990) *Organic farming*. Farming Press Books, Ipswich, England.

Lawrence, C.R. (1983) *Nitrate-rich groundwaters of Australia*, Australian Water Resources Council Technical Paper 79. Australian Government Publishing Service, Canberra.

Lawrence, G. (1987) *Capitalism and the countryside: the rural crisis in Australia*. Pluto Press, Sydney.

Lawrence, G., Vanclay, F. and Furze, B. (eds) (1992) *Agriculture, environment and society: contemporary issues for Australia*. Macmillan, Melbourne.

Lee, K.E. and Pankhurst, C.E. (1992) 'Soil organisms in sustainable productivity'. *Australian Journal of Soil Research*, 30(6): 855–92.

Leece, D.R. (ed.) (1974) *Fertilisers and the environment*. Proceedings of the symposium on ecological aspects of fertiliser technology and use. Australian Institute of Agricultural Science, University of Sydney.

Leopold, A. (1949) *A Sand County almanac*. Oxford University Press, New York.

Lewis, J. (1981) 'The economics of dryland farming'. *Quarterly Review*, 3(2): 157.

Lobry de Bruyn, L.A. and Conacher, A.J. (1990) 'The role of ants and termites in soil modification: a review'. *Australian Journal of Soil Research*, 28(1): 55–93.

Lonsdale, R.E. and Holmes, J.H. (eds) (1981) *Settlement systems in sparsely populated regions: the United States and Australia*. Pergamon, New York.

Luke, B.G., Monheit, B., Tawfik, F. and others (1988) 'Organochlorine insecticides and polychlorinated biphenyls in human breast milk, Victoria 1985/86' (cited in Scott, 1991:606).

MacArthur, R.M. and Wilson, E.O. (1967) *The theory of island biogeography*. Princeton University Press, Princeton.

McDonald, G.T. (1976) 'Rural representation in Queensland State Parliament'. *Australian Geographical Studies*, 14: 33–42.

McDonald, G.T. (1991) 'Recent developments in rural planning in Australia'. *Progress in Rural Policy and Planning*, vol. 1 (pp. 228–41) Belhaven, London.

McFarlane, D.J., Barrett-Lennard, E.G. and Setter, T.L. (1989) 'Waterlogging: a hidden constraint to crop and pasture production in southern regions of Australia'. In *Proceedings of the 5th Australian Agronomy Conference* (pp. 74–83). University of Western Australia, Perth.

McGarity, J.W. and Storrier, R.R. (1986) 'Fertilisers'. In J.S. Russell and R.F. Isbell (eds), *Australian soils: the human impact* (pp. 304–33). University of Queensland Press, St Lucia.

McGarry, D. (1990) 'Soil compaction and cotton growth on a vertisol'. *Australian Journal of Soil Research*, 28: 869–77.

McGhie, D.A. and Posner, A.M. (1980) 'Water repellence of a heavy-textured Western Australian surface soil'. *Australian Journal of Soil Research*, 18: 209–23.

McTainsh, G. and Boughton, W. (eds) (1993) *Land degradation processes in Australia*. Longman Cheshire, Melbourne.

McWilliam, J.R. (1981) 'Research and development to service a sustainable agriculture'. *Search*, 12(1/2): 15–21.

Margulis, L. and Sagan, D. (1986) *Microcosmos: four billion years of microbial evolution*. Summit Books, New York.

Marsh, B. and Carter, D. (1983) Wind erosion. *Western Australian Journal of Agriculture*, 24: 54–7.

Marshall, V.G. (1977) *Effects of manures and fertilisers on soil fauna: a review*. Commonwealth Agricultural Bureau, Slough.

Martin, P. and Lockie, S. (1993) 'Environmental information for total catchment management: incorporating local knowledge'. *Australian Geographer*, 24(1): 75–85.

Mazanec, Z. (1974) 'Influence of jarrah leaf miner on the growth of jarrah'. *Australian Forestry*, 37: 32–42.

Messer, J. and Mosley, G. (eds) (1983) *What future for Australia's arid lands?* Australian Conservation Foundation, Hawthorn, Victoria.

Minister of the Environment (1991) *The state of Canada's environment*. Minister of Supply and Services, Ottawa.

Mooney, M. (1979) *Seeds of the earth*. International Coalition for Development Action, Ottawa.

Morgan, R.P.C. (1986) *Soil erosion and conservation*. Longman, Harlow.
Mowbray, D.L. (1988) *Pesticide use in the South Pacific*, Regional Seas Reports and Studies, No. 89. United Nations Environment Programme, Nairobi.
Mues, C., Roper, H. and Ockerby, J. (1994) *Survey of Landcare and land management practices 1992–93*, ABARE Research Report 94.6. Australian Bureau of Agricultural and Resource Economics, Canberra.
Murray/Darling Basin Commission (1989) *Murray/Darling natural resources management strategy*, Working Group, Background Paper 89/1.
Murray/Darling Basin Ministerial Council (1987a) *River salinity, waterlogging and land salinisation: towards an integrated management strategy*, Technical Report No. 87/1.
Murray/Darling Basin Ministerial Council (1987b) *Murray/Darling Basin environmental resources study*. State Pollution Control Commission, Sydney.
Murray/Darling Basin Ministerial Council (1989) *Salinity and drainage strategy*. Australian Government Publishing Service, Canberra.
Myers, N. (1980) 'The problem of disappearing species: what can be done?' *Ambio*, 9: 229–35.
Nadolny, C. (1991) 'Tree clearing in Australia'. *Search*, 22(2): 43–6.
National Rangeland Management Working Group (1994) *Rangelands Issues Paper*. Australian and New Zealand Environment and Conservation Council, Agriculture and Resource Management Council of Australia and New Zealand, Canberra.
Nelson, R. and Mues, C. (1993) *Survey of Landcare and drought management practices 1991–92*. Australian Bureau of Agriculture and Resource Economics, Canberra.
Newman, P. and Cameron, I. (1982) *Attitudes to conservation and environment in Western Australia*. School of Environmental and Life Sciences, Murdoch University, Perth.
New South Wales Soil Conservation Service (1989) *Land degradation survey: New South Wales 1978–1988*.
Nix, H. (1988) 'Australia's renewable resources'. In L.H. Day and D.T. Rowland (eds), *How many more Australians? The resource and environmental conflicts* (pp. 65–76). Longman Cheshire, Melbourne.
Northcote, K.H. (1960) *A factual key for the recognition of Australian soils*. Division of Soils Report No. 4/60, CSIRO.
NRC (1989) *Alternative agriculture*. National Academy of Science Press, Washington DC.
Nulsen, R.A., Bligh, K.J., Baxter, I.N. and others (1986) 'The fate of rainfall in a mallee and heath vegetated catchment in south-western Australia'. *Australian Journal of Ecology*, 11: 361–71.
OECD (1981) *Guidelines for testing of chemicals*. Organisation for Economic Co-operation and Development, Paris.
OECD (1986) *Water pollution by fertilisers and pesticides*, 3rd edn. Organisation for Economic Co-operation and Development, Paris.
OECD (1989) *Agricultural and environmental policies: opportunities for integration*. Organisation for Economic Co-operation and Development, Paris.
Olsson, L. (1993) 'On the causes of famine—drought, desertification and market failure in the Sudan'. *Ambio*, 22(6): 395–403.

O'Riordan, T. (1976) *Environmentalism*. Pion Ltd, London.

Ormrod, D.P. (1978) *Pollution in horticulture*. Elsevier, Amsterdam.

Paoletti, M.G. and Pimentel, D. (eds) (1992) *Biotic diversity in agroecosystems*. Elsevier, Amsterdam.

Parry, M. (1990) *Climate change and world agriculture*. Earthscan, London.

Pearman, G.I. (ed.) (1988) *Greenhouse: planning for climatic change*. CSIRO, Melbourne.

Peck, A.J. and Hurle, D.H. (1973) 'Chloride balance of some farmed and forested catchments in south-western Australia'. *Water Resources Research*, 9: 648–57.

Peck, A.J. amd Williamson, D.R. (1987) 'Effects of forest clearing on ground-water'. In A.J. Peck and D.R. Williamson (eds), *Hydrology and salinity in the Collie River Basin, Western Australia*, Journal of Hydrology 94(1/2), Special Issue: 47–66.

Pigram, J.J. (1986) *Issues in the management of Australia's water resources*. Longman Cheshire, Melbourne.

Pimentel, D., Andow, D., Dyson-Hudson, R. et al. (1980) 'Environmental and social costs of pesticides: a preliminary assessment'. *Oikos*, 34: 126–40.

Pimentel, D. and Edwards, C.A. (1982) 'Pesticides and ecosystems'. *Bioscience*, 32: 595–600.

Pimentel, D. and Levitan, L. (1986) 'Pesticides: amounts applied and amounts reaching pests'. *Bioscience*, 36: 86–91.

Pimentel, D., McLaughlin, L., Zepp, A. and others (1991) 'Environmental and economic effects of reducing pesticide use'. *Bioscience*, 41(6): 402–9.

Pittock, A.B. (1989) 'The Greenhouse effect: regional climate changes and Australian agriculture'. In *Proceedings of the 5th Australian Agronomy Conference* (pp. 289–303), University of Western Australia, Perth.

Powell, J.M. (1976) *Environmental management in Australia 1788–1914*. Oxford University Press, Melbourne.

Powell, J.M. (1988) 'Patrimony of the people: the role of government in land settlement' In R.L. Heathcote (ed.), *The Australian experience* (pp. 14–24). Longman Cheshire, Melbourne.

Public Works Department (1979) *Clearing and stream salinity in the south-west of Western Australia*, Document No. MDS 1/79. Perth.

Puvaneswaran, P. and Conacher, A.J. (1983) 'Extrapolation of short-term process data to long-term landform development: a case study from south-western Australia'. *Catena*, 10: 321–37.

Raison, R.J. and Khanna, P.K. (1982) 'Modification of rainwater chemistry by tree canopies and litter layers'. In E.M. O'Loughlin and P. Cullen (eds), *Prediction in water quality* (pp. 69–86). Australian Academy of Science, Canberra.

Reeve, I.J., Lees, J.W. and Hammond, K. (1990) *Meeting the challenge: a future perspective on Australian agriculture and agricultural education*. Rural Development Centre, University of New England, Armidale.

Reid, R. and Wilson, G. (1985) *Agroforestry in Australia and New Zealand*. Goddard and Dobson, Victoria.

Resource Assessment Commission (1992) *A survey of Australia's forest resource*. Australian Government Publishing Service, Canberra.

Rickson, R.E., Saffigna, P., Vanclay, F. and McTainsh, G. (1987) 'Social bases of farmers' responses to land degradation'. In Chisholm and Dumsday (pp. 187–200).

Ride, W.D.L. and Wilson, G.R. (1982) 'The conservation status of Australian animals'. In R.H. Groves and W.D.L. Ride (eds), *Species at risk: research in Australia*, Proceedings of a symposium on the biology of rare and endangered species in Australia (pp. 27–44). Australian Academy of Science, Canberra.

Risser, J. (1985) 'Soil erosion problems in the USA'. *Desertification Bulletin*, 12: 20–5. UNEP, Nairobi.

Roberts, B. (1992) *Landcare manual.* University of New South Wales Press, Kensington.

Rose, C.W. (1988) 'Research progress on soil erosion processes and a basis for soil conservation practices'. In R. Lal (ed.), *Soil erosion research methods* (pp. 119–39). Soil and Water Conservation Society, Ankeny, Iowa.

Rose, C.W. (1989) 'Wind erosion: what research and development is still needed?' *The theory and practice of soil management for sustainable agriculture*, Workshop for Wheat Research Council, Department of Primary Industries and Energy. Australian Government Publishing Service, Canberra.

Rose, S. (1991) 'Landcare news: West Hume Landcare Group'. *Australian Journal of Soil and Water Conservation*, 4(1): 60–1.

Rudwick, V. (1991) 'Farm management characteristics'. In *Farm survey report: financial performance of Australian farms* (pp. 29–30). ABARE, Australian Government Publishing Service, Canberra.

Ryan, P.J. (1990) 'Role of trees in controlling land degradation'. In *Proceedings of the Land Degradation Conference* (pp. 47–57). Geographical Society of New South Wales, Sydney.

Schnoor, J.L. (ed.) (1992) *Fate of pesticides and chemicals in the environment.* John Wiley and Sons, New York.

Scott, D. (1991) *1991 State of the environment report. Agriculture and Victoria's environment: resource report.* Office of the Commissioner for the Environment, Government of Victoria, Melbourne.

Sedgley, R.H., Smith, R.E. and Tennant, D. (1981) 'Management of soil water budgets of recharge areas for control of salinity in south-western Australia'. *Agricultural Water Management*, 4: 313–34.

Select Committee into Land Conservation (1990) *Discussion Paper No. 2. Agricultural region of Western Australia.* Western Australia Legislative Assembly, Perth.

Select Committee into Land Conservation (1991) *Final report.* Legislative Assembly, Perth.

Senate Select Committee (1990) *Report on agricultural and veterinary chemicals in Australia.* Commonwealth of Australia, Australian Government Publishing Service, Canberra.

Sexton, R.N. (1980) 'Mobility of the Australian rural workforce'. In I.H. Burnley, R.J. Pryor and D.T. Rowland (eds), *Mobility and community change in Australia* (pp. 67–78). University of Queensland Press, Brisbane.

Seymour, J. (1980) 'Soil erosion: can we dam the flood?' *Ecos*: 25, 3–11.

Shearer, B.L. and Tippett, J.T. (1989) *Jarrah dieback: the dynamics and manage-*

ment of Phytophthora cinnamomi *in the jarrah* (Eucalyptus marginata) *forest of south-western Australia*, Research Bulletin No. 3. Department of Conservation and Land Management, Como, Western Australia.

Simpson, J. and Curnow, B. (eds) (1988) *Cadmium accumulations in Australian agriculture*, National Symposium, Bureau of Rural Resources. Australian Government Publishing Service, Canberra.

SIPRI (1990) *World armaments and disarmament*, SIPRI Yearbook. Oxford University Press, Oxford (cited in Tolba et al. (1992).

Skarby, L. and Sellden, G. (1984) 'The effects of ozone on crops and forests'. *Ambio*, 13(2): 68–72.

Smil, V. (1992) 'China's environment in the 1980s: some critical changes'. *Ambio*, 21: 431–36.

Smith, D.D. and Wischmeier, W.H. (1962) 'Rainfall erosion'. *Advances in Agronomy*, 14: 109–48.

Specht, R.L. (1981) 'Conservation of vegetation types'. In R.H. Groves (ed.), *Australian vegetation* (pp. 393–410). Cambridge University Press, Cambridge.

Specht, R.L., Roe, E.M. and Boughton, V.H. (1974) 'Conservation of major plant communities in Australia and Papua New Guinea'. *Australian Journal of Botany*, Supplementary Series No. 7.

State Pollution Control Commission (1980) *Namoi environmental study*. New South Wales Department of Agriculture, National Parks and Wildlife Service and Others, New South Wales.

Swarbrick, J.T. (1986) 'Pesticides'. In J.S. Russell and R.F. Isbell (eds), *Australian soils: the human impact* (pp. 357–73), University of Queensland Press, St Lucia.

Taylor, M.J. (1991) 'Economic restructuring and regional change in Australia'. *Australian Geographical Studies*, 29: 255–67.

Tiller, K.G. (1989) 'Heavy metals in soils and their environmental significance'. *Advances in Soil Science*, 9: 113–42.

Tolba, M.K., El-Kholy, O.A., El-Hinnawi and others (1992) *The world environment 1972–1992: two decades of challenge*. Chapman and Hall on behalf of the United Nations Environment Program, London.

Tomkins, I.B., Kellas, J.D., Tolhurst, K.G. and Oswin, D.A. (1991) 'Effects of fire intensity on soil chemistry in a Eucalypt forest'. *Australian Journal of Soil Research*, 29: 25–47.

Toyne, P. and Farley, R. (1989) 'A national land management program: joint ACF/NFF submission'. *Australian Journal of Soil and Water Conservation*, 2(2): 6–7.

Trotman, C.H. (ed.) (1974) *The influence of land use on stream salinity in the Manjimup area, Western Australia*, Technical Bulletin No. 27. Department of Agriculture of Western Australia, Perth.

UNDP (1991) *Human Development Report 1991*, United Nations Development Program. Oxford University Press, Oxford.

UNEP (1989) *Environmental Effects Panel Report*. United Nations Environment Programme, Nairobi.

UNEP (1992) *World atlas of desertification*. Edward Arnold, London.

UNEP (1993) *World map of present-day landscapes: an explanatory note*. United Nations Environment Programme and Moscow State University, Moscow.

United States Department of Agriculture (1980) *Report and recommendations on organic farming*. Beltsville, Maryland.

Valentine, P.S. (1976) A *preliminary investigation into the effects of clear-cutting and burning on selected soil properties in the Pemberton area of Western Australia*, Geowest No. 8. Department of Geography, University of Western Australia, Perth.

Vanclay, F. (1992) 'The social context of farmers' adoption of environmentally sound farming practices'. In Lawrence et al. (pp. 94–121).

Wace, N. (1988) 'Naturalised plants in the Australian landscape'. In R.L. Heathcote (ed), *The Australian experience: essays in land settlement and resource management* (pp. 139–50). Longman-Cheshire, Melbourne.

Wallace, M.M.H. (1970) 'The biology of the jarrah leaf miner, *Perthida glyphopa* Common (Lepidoptera: Incurvariidae)'. *Australian Journal of Zoology*, 18: 91–104.

Warner, R.F. (1984) 'Man's impacts on Australian drainage systems'. *Australian Geographer*, 16, 133–41.

Wasson, R.J. (1990) 'Greenhouse warming and land degradation in New South Wales'. In *Proceedings of the Land Degradation Conference* (pp. 59–70). Geographical Society of New South Wales, Sydney.

Wasson, R.J. and Galloway, R.W. (1984) 'Erosion rates near Broken Hill before and after European settlement'. In R.J. Loughran (comp.), *Drainage basin erosion and sedimentation* (pp. 213–19). University of Newcastle and New South Wales Soil Conservation Service, Newcastle.

WAWRC (1992) *The state of the rivers of the southwest*, Publication No. WRC 2/92, Western Australian Water Resources Council, Perth.

Wells, K.F., Hood, N.H. and Laut, P. (1984) *Loss of forests and woodlands in Australia: a summary by State, based on rural and local government areas*. CSIRO, Canberra.

WHO (1990) *Public health impact of pesticides used in agriculture*. World Health Organisation, Geneva.

Willat, S.T. and Pullar, D.M. (1983) 'Changes in soil physical properties under grazed pastures'. *Australian Journal of Soil Research*, 22: 343–8.

Wild, A. (1993) *Soils and the environment: an introduction*. Cambridge University Press, Cambridge.

Williams, W.D. (ed.) (1980) *An ecological basis for water resource management*. Australian National University Press, Canberra.

Williamson, D.R. (1983) The application of salt and water balances to quantify causes of the dryland salinity problem in Victoria'. *Proceedings of the Royal Society of Victoria*, 95(3): 103–11.

Williamson, D.R. and Bettenay, E. (1979) 'Agricultural land use and its effect on catchment output of salt and water—evidence from southern Australia'. *Progress in Water Technology*, 11: 463–80.

Williamson, D.R., Stokes, R.A. and Ruprecht, J.K. (1987) Response of input and output of water and chloride to clearing for agriculture. In A.J. Peck and D.R. Williamson (eds), *Hydrology and salinity in the Collie River Basin, Western Australia, Journal of Hydrology*, 94: 1–28.

Wills, R.T. (1993) 'The ecological impact of *Phytophthora cinnamomi* in the Stirling Range National Park, Western Australia'. *Australian Journal of Ecology*, 18: 145–59.

Wischmeier, W.H. (1976) 'Use and misuse of the universal soil loss equation'. *Journal of Soil and Water Conservation*, 31: 5–9.

Wischmeier, W.H. and Smith, D.D. (1978) *Predicting rainfall erosion losses*, Agriculture Handbook No. 537, Science and Education Administration. US Department of Agriculture, Washington DC.

Wolfe, S.A. and Nickling, W.G. (1993) 'The protective role of sparse vegetation in wind erosion'. *Progress in Physical Geography*, 17: 50–68.

Woods, L.E. (1983) *Land degradation in Australia*. Australian Government Publishing Service, Canberra.

Woods, W. (1981) 'Controlling cotton pests with egg parasites'. *Western Australian Journal of Agriculture*, 22: 63–4.

World Commission on Environment and Development (1987) *Our common future* (The Brundtland Report). Oxford University Press, Oxford.

World Food Council (1991) *Hunger and malnutrition in the world*, World Food Council Doc. WFC/1991/2. WFC, Rome.

Young, M.D. (1991) *Towards sustainable development*. OECD, Belhaven Press, London.

Index

Note: Locators in bold indicate relatively major textual references. Locators followed by an (*) indicate figures, those followed by a (t) indicate tables and those followed by a (p) indicate plates.

ABARE (Australian Bureau of Agricultural and Resource Economics) 65, 107, 108, 109, 110, 111, 144
Aborigines 3, 86
Abrahams, A.D. 95, 144
ABS (Australian Bureau of Statistics) 21, 25, 26, 27, 30, 49, 60, 61, 62, 63, 66, 75, 77, 78, 86, 110, 144
ACT (*see* Australian Capital Territory)
Africa 6, 8, 9
agricultural areas, land 4, 13, 16, 18, 40, 64, 73, 78
  cereals 39, 66t
  clearing 10, 12, 67, 73, 75
  crops 106, **106***, 109
  dairying 44, 66t, 108, 119, 138
  expansion **65–7**, 75
  extensive cropping (broadacre) 4, 14, 47, 50*, 65, 108, 119, 138
  hobby farming 76
  horticulture 4, 39, 41, 44, 65, 66t, 72t, 109, 129
  irrigated lands 5, 10, 11, 43, 67–8, 73, 73*, 136
    Camballin irrigation scheme 67
    costs **67–8**
    Humpty Doo irrigation scheme 67
    Namoi cotton area 52
    Ord irrigation scheme **67–8**
    Tipperary irrigation scheme 67
  livestock farming 4, 47, 66t, 106, 108, 129
  market gardens (*see* horticulture)
  pastoral areas, country, land, rangeland 4, 13, 18, 121, 126
  pastures 38, 39, 40, 79t, 106, **106***
  piggeries 44, 66t
  poultry 66t
agricultural chemicals (*see also* veterinary chemicals, pesticides, synthetic fertilisers) **38–57, 66t,** 111
agricultural practices
  agricultural machinery 49, 110, 111*, 131, 132
    tractor sales 110, **111***
  crop rotations 111
  cultivation 49, **100–1**
    effects on soil **100–1**
    effects on erosion 101–2, **103***
  irrigation practices 71–2
  stocking
    effects on soil, erosion 101–2

agriculture — role in Australian economy **106–7**
air pollution (*see* pollution)
alfalfa 72t, 79t, 91t
algae (*see* eutrophication)
ANOP (Australian National Opinion Polls) 110, 114, 115, 145
aquifer (*see* water movements)
Armillaria 85
Asia 9
Australia
  arid zone 13*
  land degradation — extent, nature of 6, 13*, 15t, **12–18**, 20, 20t, 42, 53, 58, 59t
  land use 13*, 34–5t
  Prime Minister's Statement on the Environment 1989 12, 138, 143
  1975–77 survey of land degradation 4, 13*, 15t, **12–18**, 19, 24, 58, 59t, 60, 62,
  1989 federal inquiry 12
Australian and New Zealand Environment and Conservation Council (ANZECC) 26–31
Australian Bureau of Agricultural and Resource Economics (*see* ABARE)
Australian Bureau of Statistics (*see* ABS)
Australian Capital Territory 13*, 15t, 20, 20t, 21t, 59t, 60t, 125t, 136, 139*
  Canberra 43
  Lower Molonglo 43
Australian Conservation Foundation 138
Australian, The 70
Australian Journal of Soil and Water Conservation 125
Australian National Opinion Polls (*see* ANOP)
Australian Year Book 21

Baghdad 10
barley 72t
beans 72t, 79t
beets 72t
Benson, J. 31, 145
Billion Trees Program 22
biophysical processes **83**
  nutrient cycling 84, 135t
  translocation of soil materials 92, 94
biophysical properties 83
Blyth, M.J. 62, 121, 152
Bolton, G. 32, 112, 145
Boon, S. 86, 145
Bowie, I.J.S. 81, 82, 145
Breckwoldt, R. 127, 145
Briggs, J.D. 26, 27, 145
Brookfield, H. 10, 145
Brown, A.W. 46, 145
Brundtland report 139
Bryan, R.B. 95, 146
Bureau of Agricultural Economics 82, 121
Bureau of Rural Resources 36, 37, 47, 66
Butcher, E. 4, 125, 136, 137, 146

Cameron, I. 114, 115, 154
Campbell, A. 81, 117, 138, 139, 146
Canada 25, 26, 31, 33, 66t, 122
  Ontario 129
Carne, R.J. 135, 146
Central America 6, 10, 12
cereals 39
China 6, 10, 11
  Bohai Bay 11–12
  coastal plain 11
  Gobi 11
  Henan Province 12
  Hwang He (Yellow River) 11
  Jen Chou valley 11p
  Jinan 12
  Lanzhou 11
  loess plateau 10, 11, 11p
    gullying 11, 11p
    landslides 11, 11p
    subsurface tunnel-gully erosion 11
  North China Plain 12
  Shandong Province 12
climate change, variability 8, 10, 77–8
  agricultural implications 10, **77–8**
  El Niño **78**
  greenhouse effects 2, 77, 78, 140
clover 72t, 79t
coastal zone 4
Commission on Plant Genetic Resources 128
Commonwealth Department of Health Poison Centre 54
Commonwealth Department of Primary Industries and Energy 33, 35, 64, 129
Commonwealth of Australia 140, 147
Conacher, A.J. 3, 47, 49, 50, 51, 52, 72, 89, 92, 97, 104, 116, 118, 120, 121, 130, 131, 133, 134, 144, 147, 150, 153, 155
Conacher, J.L. 47,49, 50, 51, 52, 118, 120, 121, 130, 131, 133, 134, 147
conservation (*see also* land degradation solutions, soil conservation) 116, 135
conservation reserves (*see* land degradation solutions, reserves)
Consumer Price Index (CPI) 62
corn 72t, 79t, 91t
cost/price squeeze 67, 80

cotton 65, 66t, 67, 72t, 79t, 101, 130
country towns (*see also* population) 82
  decline **80–1**
  social changes 81
Cribb, A.B. 31, 147
Cribb, J. 39, 42, 106, 107, 108, 117, 122, 144, 147
Cribb, J.W. 31, 147
crop pests and diseases 36–7, 65, 101, 131, 132
  aphids 36
  bacteria 36
  fungi 36, 84–5
  mites 36
  molluscs 36
  nematodes 36
  viruses 36
CSIRO 41, 52, 70, 147

Darling system (river) 43
Davidson, R. 25, 127, 148
Davidson, S. 25, 127, 148
Decade of Landcare 138
Decade of Landcare Plan, WA 136
deforestation 6, 10, 12
degradation (*see* land degradation)
Department of Agriculture, Vic. 54
Department of Agriculture, WA 16, 64, 119, 120
Department of Arts, Heritage and Environment 12, 14–16, 19, 20, 21, 23, 25, 26, 32, 70, 71, 148
Department of Conservation and Land Management, WA 26, 27t
Department of Environment, Housing and Community Development 12, 24, 148
Department of Water Resources, NSW 97
De Ploey, J. 95, 148
desertification (*see also* land degradation) 3, 4, 6, 8, 10
  economic costs 8t, 9t
  manifestations 8–9
  myth 8
direct drilling (*see also* minimum tillage) 49
  advantages 50–1
  disadvantages (*see under* pesticides)
Dodson, J.R. 86, 145
DPI (Department of Primary Industry) 46, 148
DPIE/ASCC (Department of Primary Industries and Energy/Australian Soil Conservation Council) 121, 148
Dragovich, D. 115, 141, 148
Dregne, H.E. 8, 11, 148, 149
drought 13, 106, 108, 114p, 118, 135t
drylands 6
Dunlap, R.E. 115, 149
dykes 12

Eastern Europe 9
ecologically sustainable development (ESD) **140**, 149
  Brundtland report 139
  Compendium of ESD Recommendations 140
  definition **140**
  National Strategy 140, 143
ecosystems
  degradation 2, **19–37**
  ecological disturbance 45, 86
  genetic resources loss **31–2**, 84
  habitat destruction 84, 125
  introduced pests 36–7
  introduced species 45, 84
  nutrient cycling **86–9**, 135t
  Specht survey of conservation status 22, 23t, 24
  species loss 25–31, 84
Egypt 10
Ellington, T. 64, 149
El-Swaify, S.A. 124, 149
Emmett, W.W. 95, 149
England 44, 79
ESD (*see* ecologically sustainable development)
Euphrates 10
Europe 6, 9, 31, 44, 53, 79
European Community 44, 107, 122
European settlement (of Australia) 19, 20, 31, 32, 33, 73, 75, 86, 96, 116
eutrophication (*see also* water) **41–3**, 68, 69, 70, 77, 136
  algae, algal blooms 42, 43*, 68, 70
  nutrient sources 42, 43, 69
evaporation (*see under* water)
evapotranspiration (*see under* water)
extinct, rare and endangered species **25–31**

FAO (Food and Agriculture Organisation of the United Nations) 8, 9, 53, 63, 128, 149
famine 8–9
Farley, R. 111, 138, 142, 149, 157
farms
  costs 49, 49*, 107
  deferred farm expenditures 110, 111
  efficiency 106
  hobby farms 76, 82

farms (*cont.*)
  increased area per agricultural worker 110
  increased farming intensity **69**
  labour (*see* rural workforce)
  productivity 106, 111
  reduced number 80, 109, **109***
  sizes, increase 66–7, 80, 109
fauna
  amphibians 25, 26, 28t, 32
  birds 25, 26, 28–9t, 32
  extinctions, rates **25–31**
  freshwater fish 25, 26, 30t, 32
  insects 25, 64
  mammals 25, 26, 29–30t, 32
  molluscs 25
  reptiles 25, 26, 30t, 32
  vertebrates 25
feral animals (*see also* introduced species) 32, 33
fertilisers (*see* synthetic fertilisers)
fire
  ecosystem stress **86**
  frequencies 86
  prescribed (controlled) burning **86**, 89
flooding, floods 8, 12, 69
flora
  endangered 27t
  extinctions, rates 27t, **25–31**
  introduced plants 32
  vascular plants 25
forests (*see also* vegetation)
  clearing **19–22**, 90*, 93, 97
  coniferous 91t
  cypress 21t
  deciduous 91t
  definition 20
  Eucalyptus 20, 21t, 26, 90, 94
  extent 19, 21, 21t, 23t, 75
  fire 84, **86**, 98p
  jarrah 24, 86, 92, 97
  karri 87
  litter, litter layer 86, 89, 92, 98p
  litter dams **98p, 99p**
  logging 84
  losses 20, 20t, 22, 24, 79
  plantations 20, 21, 21t
    coniferous 75
    eucalyptus 75
  rainforests 20, 21, 21t
  replanting, reforestation 22
  stress 84
  tropical 21t
  wandoo 98–9p
Foster, G.R. 95, 149
Fry, P.W. 119, 149

Galloway, R.W. 96, 97, 158
Garman, D.E.J. 70, 150
George, R.J. 14, 90, 94, 150
Girsu 10
Goss, K.F. 119, 149
government agencies, policies (*see under* land degradation, underlying causes)
Government of Canada 25, 26, 33, 66, 124, 150
Graetz, D. 1, 20, 125, 150
Grant, R. 4, 17, 63, 78, 101, 150
grass 72t, 79t
Greening Australia 22
groundwater (*see* water movements)
  groundwater table rise **97, 100**
Guenzi, W.D. 46, 150
Gutteridge Haskins and Davey 42, 150

Hansard 47
Hawke, R.J.L. 1, 12, 138, 151
Heathcote, R.L. 112, 116, 145, 147, 151, 155, 158
heavy metals **40–1**
  cadmium 41, 56t
  health implications **41**, 56t, 77
Hellden, U. 8, 151
herbland 23t
House of Representatives Standing Committee on Environment 12, 62, 64, 100, 151
Hudson, N. 124, 151
Hussey, B.M.J. 25, 84, 86, 127, 151

ICI (Imperial Chemical Industries) 49
ICM (integrated catchment management: *see under* land degradation, solutions)
India 31
Indus valley 5
integrated catchment management (ICM) (*see also* total catchment management (TCM), *under* land degradation, solutions)
International Board for Plant Genetic Resources (IBPGR) 128
International Undertaking on Plant Genetic Resources 128
Iraq 10
IUCN/WWF (International Union for the Conservation of Nature/World Wildlife Fund) 32, 151

Japan 67, 77, 122
jarrah leaf miner (*Perthida glyphopa*) 24, 25
jarrah (*Eucalyptus marginata*) 24, 86
Jeans, D.N. 112, 147, 151
Joint Committee 68, 152

kangaroos 37
Kellehear, A. 80, 152
Kirby, M.G. 62, 121, 152

land, definition 2
Land and Water Resources Council 142
Land and Water Resources Research and Development Corporation (LWRRDC) 61t
Landcare (*see under* land degradation, solutions)
Land Conservation Districts (*see also* Soil Conservation Districts) 32, 119
land degradation (*see also* desertification, soil degradation)
  Australia (states listed separately) 6, **12–18**, 58, 59t, 64
  causes — direct **83–103**
  coastal zones 4
  costs 36, 37, 64, **58–68**
    of repair 8, 9t, 59t, 60t, **58–62**
  crop losses 64, 78, 79t
  definition **2–3**, 55
  deforestation 10, 12, **19–22**
  diseases (crops, animals) 8, 24, 36–7, 45, 65, 68, 84, 101
  ecological disturbance, effects **24–5**, 45, **84–5**, 86
  economic implications **8, 58–68**
  ecosystems **19–37**
  expenditure 60, **60t**
  geographers' interest, roles **5–6**
  global **6–12**
  goats 10p, 32
  habitat destruction, loss 24, 84
  history **10**
  human implications **8**
    diseases 9
    famine 9
    food 9
    malnutrition 9
  introduced species 4, **32–3**, 32, 33
  invasion by woody shrubs 16, 17t
  loss of arable land 8, 76
  overgrazing 10
  off-farm degradation **69**
  population pressures 10, 12
  production losses 36, 37, 59, **62–5**, 69
  productivity static, declining 62, 63
  salinity (*see* salinisation)
  sedimentation 8, 10, 12, 68, 69, 70, 77, 102–3, 136
  social implications **79–82**
  solutions **124–43**
    agroforestry 132, **134–5**, 135t
    biodynamic farming 130
    biological control 52
    clearing bans 22
    conservation farming 69
    controlled grazing 127, **133**
    deep ripping **133**
    ecologically sustainable land management 133–4, **139–40**
    education 140, 142–3
    grade banks 120, **132**
    gypsum applications **133**
    integrated catchment management (ICM) 4, 11p, 134, **136–7**, 140
    Integrated Pest Management (IPM) **129–30**, 140
    Landcare 60, 61t, 111, 119, 134, **137–9**, 139*, 140, 142, 143
    legumes 132
    liming 133
    modified rotation practices **132**
    organic farming 130–1, **133–4**
    planting trees, shrubs 11p, 22, 127, 134–5, 138, 143
    protecting genetic diversity 127, 128
    rehabilitation of ecosystems 127
    reduced tillage 132
    reduced agricultural chemicals **128–31**, 133
    reserves 75, 125t, **125–6**
      requirements **125–6, 126***
    soil conservation 60, 60t, 61t, 78, 111, 114, 118, 119, 120, 121, 130
    Soil Conservation Districts (now termed Land Conservation Districts) **137–9**
    stubble retention 131
    sustainable agriculture 133–4, 141, 142
    terracing 11p
    total catchment management (TCM: *see* integrated catchment management)
    whole farm management 4, 134, **136**
    who pays? **141**
    windbreaks **132**
  species loss 4, **25–31**, 84
  underlying causes **104–22**
    attitudes 104, 105, 112, 115t, **113–17**, 122–3, 143
      Judaeo-Christian attitude — 'domination of nature' 113, 116
    'battlers' 116
    conservatism 116, 117, 118

land degradation (*cont.*)
economic constraints, context, pressures 104, **105–111**, 143
cost/price squeeze 107, **107***, 110
farm debt **108, 108***
terms of trade **107**
education **117–18**, 119
government agencies 105, **119–21**, 137
government agency structures 105, 119, 142
government legislation 105, 121, 142
government policies 105, **121–2**
drought relief 121
price supports 122
subsidies 121–2
tax concessions 121–2
grandiose schemes 122
leasehold 121
lack of awareness 104, 105, **118–19**
multinationals 104–5
nature of the biophysical environment 105, **112–13**
aridity 112
lack of fresh water 112
poor soil fertility 89, 112
unfamiliar vegetation 112
perceptions 104, 105, 112
pioneering mentality 104, **116–17**
political influence 105, 116–17, 121
recency of settlement by Europeans 104, 116
resistance to change 105, **117–18**
rural ethos **116–17**
rural sociology 117
urban areas 4, 77
vegetation degradation, decline, clearing 4, 14, 15t, 18*, 20t, 24, 67, 75, 84, 126
yield declines 62, 63*, 100, 101
land management practices **83**, 93
land-use conflict **75–9**
indirect **76–9**
Latin America 9
Lee, K.E. 89, 152
loess 11
Lobry de Bruyn, L.A. 88–9, 118, 153

McGarry, D. 101, 153
Mackenzie 118
McWilliam, J.R. 62, 153
mallee fowl 86
Mayan culture 12
MED (International Geographical Union Study Group on Desertification in Regions of Mediterranean Climate) 3
MEDALUS (Mediterranean Desertification and Land Use) 3
Mediterranean 10
climate 63, 132
Mesopotamia 5, 10, 37
Messer, J. 4, 153
methaemoglobinaemia ('blue baby syndrome') 45
Mexico 12
Middle Ages 10
Middle East 6
minimum tillage (*see also* direct drilling) **48–52**, 65, 131
Mitchell grasslands 24
Mongolia 6
Morgan, R.P.C. 124, 154
Mosley, G. 4, 153
Murray/Darling Basin Commission 138
Murray/Darling Basin Natural Resources Management Strategy 136–7
Murray/Darling drainage basin 16, 42, 43*, 43, 67, 68, 70, 97, 136
Murray river 43*, 70, 72, 73, 73*, 74*
Murrumbidgee river 43*, 43, 70, 72

National Farmers' Federation 111, 116, 138, 142
National Reafforestation Program 138
National Resources Development Research Corporation 138
National Soil Conservation Program (NSCP) 61t, 138, 142, 143
National Strategy for Ecologically Sustainable Development 140, 143
National Weeds Strategy 33, 140
natural hazards 3
natural predators 24–5
New South Wales 13*, 14, 15t, 16, 17t, 20, 20t, 21, 21t, 22, 23t, 24, 26, 27t, 36, 40, 43*, 44, 59t, 60, 60t, 62, 70, 81, 96, 105, 115, 116, 120, 125t, 129, 132, 136, 137, 139*, 142
Albury 43, 97
Broken Hill 96
Carcoar reservoirs 42
Catchment Management Act 136
Hay weir pool 42
Namoi cotton area 52
Soil Conservation Service 16, 154
Sydney 36
Wagga Wagga 43
Newman, P. 114, 115, 154
New Zealand 117

nitrates (*see also* pollution) **44–5**, 69
Nix, H. 66, 154
non-arid grazing areas 14
North Africa 5
North America 6, 9, 31, 44, 62, 63*
Northcote, K.H. 93, 154
Northern Territory 13*, 15t, 20, 20t, 21t, 23t, 27t, 33, 59t, 60t, 67, 116, 120, 125t, 138, 139*
  Arnhem Land 36
  Darwin 85
  Humpty Doo irrigation scheme 67
  Kakadu National Park 36
  Tipperary irrigation scheme 67
noxious weeds 33
NRC (National Research Council) 53, 130, 134, 154
Nulsen, R.A. 91, 154
nutrients
  calcium 88t
  cycling **86–9**
  excess nutrients — toxic excesses 40
  induced deficiencies 39, **40**
  loss from agricultural systems 41–2, 70, 89
  magnesium 88t
  nitrogen 88t, 89, 132
  phosphorus 88t, 89
  potassium 88t, 89

oats 91t
OECD (Organisation for Economic Co-operation and Development) 8, 39, 46, 53, 62, 63, 69, 77, 79, 106, 154
Olsson, L. 8, 155
One Billion Trees Program 138
O'Riordan, T. 45, 113, 155
ozone (*see under* pollution)

Pankhurst, C.E. 89, 152
parasites 36–7, 129
  ticks 37
  worms 36
pedogenesis (*see* soil formation)
pesticides 4, 131
  adverse effects **45–57**, 65, 68
  benefits 64, 65
  contamination of water 53, 54, 70, 71t
  costs 36, 37, 52, 58, **65, 66t**
  DDT 52, 53, 56t, 67, 68
  dieldrin 53, 56t
  fungicides 36, 55, 65
  herbicides (*see also* minimum tillage) 47, 47*, 48, 49, 55, 65
    areas treated 49, **50***
    environmental effects 51
    health effects 54, 57*
    indirect costs 51
    paraquat/diquat 49
  insecticides 36, 55, 65
  in the environment **48**, 52, 53, 65, 68
  organochlorines 48, 53
  pest resistance 52, 64, 65, 67
  poisoning 46, **54–7**, 65
    symptoms 46, 54, 57*
  residues in food 53, 54, 68
    human health 53, **54–7**, 55*, 56–7t, 57*
  restrictions and cancellations 56–7t, **129–30**
  spray drift **52–3**, 54
    ecosystem damage 53, 54
  standards 53
  synergistic effects 46–7
  toxicity **46–7**, 48
    LD (Lethal Dose) 46
  use in Australia 47, 47*
pests (*see also* crop pests and diseases) 4, 36–7, 45, 64, 65
Phytophthora cinnamomi **84–5**
Pimentel, D. 45, 47, 52, 53, 54, 64, 65, 84, 129, 149, 152, 155
plantations (*see* forests)
Plant Production Committee 33
'polluter pays' principle 141
pollution (*see also* pesticides, synthetic fertilisers, veterinary chemicals)
  nitrates 39, **44–5**, 69, 71t
    drinking water standards 44
    health implications **44–5**
  ultra-violet radiation (hole in ozone layer) 2, **78–9**
    agricultural implications **78–9**
  urban and industrial 44, 77
    acid precipitation 77
    effects on agriculture 77
    dioxins 77
    fluoride emissions 77
    heavy metals 70, 77
    nitrogen oxides 77
    ozone, excess 77, 79, 79t
    photochemical smog 79
    sewage, septic tanks 43, 44, 69
population (*see also* country towns)
  growth, pressure 5, 6, 8, 11, 12, 109
  rural population decline **79–80**, 81
  rural–urban drift 80
potato 72t, 79t
Powell, J.M. 68, 75, 112, 155
precautionary principle **45**, 52

Public Works Department 76, 155
Puvaneswaran, P. 92, 97, 155

Queensland 13*, 14, 15t, 20, 20t, 21, 21t, 22, 23t, 24, 26, 27t, 33, 36, 44, 59t, 60, 60t, 67, 70, 81, 85, 96, 108, 116, 118, 125t, 132, 136, 137, 139*
  Capricornia 16
  Darling Downs 16
  Great Barrier Reef 70
  North Pipe dam 42
  South Burnett 16
  Toowoomba 43

rabbits 32, 37
rainforest (*see* forests)
rangelands (*see also* pastoralism, pastoral country) 4
Reeve, I.J. 120, 155
remote sensing 8
Resource Assessment Commission 21, 22, 156
rice 31, 72t, 77
Roberts, B. 125, 136, 137, 156
Rose, C.W. 95, 102, 156
ROTAP (Rare or Threatened Australian Plants) 27t
Rudwick, V. 118, 156
rural debt 82
rural workforce 79–81, 109, **110***
rye 91t

Sahara 6, 8, 9
Sahel 8
salinisation, salinity 4, 5, 10, 13, 14, 37, 70, 136
  control 135t
  dryland 14, 15t, 18*, 64
  ecological effects 70
  irrigation area salinity 14, 15t, 17t, 37, 71, 72
  primary salinity 14, **100**
  salt tolerance, susceptibility 70, 71, **72t**, 75
  secondary salinity 14, 25, **100**, 118
  water salinity **70–5**, 74*, 75–6, 77
Sand Drift Act (WA, SA) 120
Save the Bush 138
Schnoor, J.L. 46, 156
Scott, D. 16, 22, 26, 32, 39, 42, 44, 53, 54, 57, 76, 78, 89, 102, 103, 153, 156
Sedgley, R.H. 94, 156
sedimentation
Select Committee into Land Conservation 18, 78, 156
Senate Inquiry into Agricultural Chemicals 47, 53, 129, 134, 140
Senate Select Committee 36, 47, 51, 52, 53, 64, 65, 129, 130, 134, 156
shrubland 23t, 24, 36
SIRATAC 130
Smailes, P.J. 81, 82, 145
soil
  A horizon 87*, 92
  clay content 93, 133
  decomposition (of litter) 88
  duplex profile 93
  hard setting 92
  horizons 87*, 92, 94
  litter layer 86, 89, 92
    water storage 92, 94
  mixing — soil biota 88, 92
  nutrients 86–7, 88t
  organic matter content 92, 118, 131, 132
  pedon 86, 87*, 93, 94*
  permeability 101
  pH 40, 41, 48, 71t, 77, 118–19, 133
  porosity 93, 97
  properties 86, 100–1
  texture 92, 100
Soil and Water Conservation Association of Australia 124
soil conservation 60, 60t, 61t, 118
  first national conference 120
  legislation 120
Soil Conservation Society of America 124
soil degradation 4, 6, 7*
  acidity 13, 16, 17t, 17, 18*, **39–40**, 63, 64, 118–19, 133, 137
  anaerobic conditions 100
  chemical soil degradation 6
  compaction 6, 17, 18*, 64
  erosion (*see* soil erosion) 5
  excess sodium 133
  hardpans 16, 64, 101, 131
  heavy metal residues 16, 39, **40–1**
    cadmium **41**, 56t
  nutrient loss, deficiencies 16, 39, **40**, 90*, 133
  nutrients — excess 16, **40**
  organic matter decline 16, 90*
  organisms **41**
  pesticide residues 16
  plough pans 16, 101
  salinity (*see also* salinisation) 5, 6, 10, 13, 14, 15t, 17t.18*, 25, 64, 68, 97, 120
  scalding 16, 17t

soil degradation (*cont.*)
sealing 95, 97, 101
soil biota decline 16, 39, 41
soil structure decline, loss 13, 16, 17t, 17, 18*, 64, 68, 100, 101, 133, 137
subsidence 6
trace element deficiences 39
trafficability 100
waterlogging 6, 10, 16, 17, 18*, 64, 68, 97, 100, 101
water repellency 16, 17, 18*, 39, 63, 64, 92, 97, 137
soil erosion
accelerated erosion 11, 12, **95–7**, 120
by water 4, 6, 11, 14, 15t, 16, 17t, 17, 18*, 39, 95, 98–9p, 101, 118, 131, 132, 135t, 136 by wind 4, 6, 14, 15t, 16, 17t, 17, 18*, 62–3, 69, 101p, **101–3**, 131, 135t, 137
fines, dust, nutrient loss 62, 102
combined wind and water erosion 14, 15t, 18*
geomorphic erosion 95
gullying 11, 17t
mass movement 11, 16, 17t
rates **96–7**
reduced erosion rates 96, 97, 131
soil loss tolerances 95
Universal Soil Loss Equation (USLE) 95
soil formation, genesis 93
rates **96**
soil salinity (*see* salinisation, soil degradation)
soil water 93
infiltration 39, 92
leaching 88, 92, 93
Somalia 9
sorghum 72t
South Asia 31
South Australia 13*, 15t, 16, 20t, 21, 21t, 23t, 24, 27t, 44, 59t, 60, 60t, 70, 75, 108, 120, 125t, 136, 139*
Adelaide 73
Murray river 43, 73, 74*
Mt Bold 42
Mt Gambier 44
South America 6, 9
southern Africa 6
Soviet Union 9, 37
soybean 79t, 91t
Spain 10p
Specht, R.L. 22, 23, 24, 157
species (*see also* ecosystems)
exotics 84
gene pools 84
introduced species **32–3**, 140
loss, extinctions **25–31**, 84
rare and endangered **25–31**
spinifex grasslands 24
Spray.Seed 49
Standing Committee on Agriculture 33
State Pollution Control Commission, NSW 52, 157
sugar cane 65, 66t, 67, 118
superphosphate (*see also* phosphate, phosphorus) 41
sustainable development 139
Sydney 36
synthetic chemicals 4, 38–57
synthetic fertilisers
adverse effects **38–45**
applications 38–9, 39t
impurities 40
nitrogen 38, 39t, 42, 43, 44, 64, 131
phosphate, phosphorus 38, 39t, 40, 42–3,
potassium 39t, 40
lime 40
use 39t
synthetic fertilisers and soil organisms **41**

Tasmania 13*, 14, 15t, 20t, 21, 21t, 23t, 27t, 59t, 60t, 70, 120, 125t, 138, 139*
Taylor, M.J. 80, 81, 157
TCM (total catchment management: *see* integrated catchment management (ICM) *under* land degradation, solutions)
Tigris 10
Tiller, K.G. 40, 41, 77, 157
Tolba, M.K. 6, 8, 31, 38, 52, 68, 78, 79, 128, 157
total catchment management (TCM) (*see* integrated catchment management (ICM))

UNDP (United Nations Development Programme) 8, 157
UNEP (United Nations Environment Programme) 2, 3, 6, 8, 78, 157–8
United Nations 8
Environment Programme (*see* UNEP)
Food and Agriculture Organisation (*see* FAO)
United States 53, 56–7t, 65, 77, 81, 94, 95, 106, 107, 122, 129, 141
California 44
Department of Food and Agriculture 55, 146
Department of Agriculture 120, 122, 133, 158

United States (*cont.*)
  Iowa 53
  Ohio 53
  pests and weeds, production losses 64
urban expansion 76, 77
'user pays' principle 141
UV-B radiation (*see* pollution, ultra-violet radiation)

vegetation (*see also* forests, rainforests, woodlands, plantations)
  arid zone 24
  bluebush 24
  canopy 89, 91t
  clearing 20t, 19–22, 67, 75, 84, 89, 97, 113, 114p
  degradation 22–4, 79
  effects on wind erosion **101, 101p, 102***
  grassland communities 23t, 22–4
  Mitchell grasslands 24, 85
  mulga 24
  replacing native vegetation with introduced species 93, 97, 101
  representation in reserves 23t, 24, 126
  saltbush 24
veterinary chemicals 36, 47*, **47–8**, 65
  in the environment 48
Victoria 13*, 14, 15t, 16, 20, 20t, 21, 21t, 22, 23t, 24, 26, 27t, 32, 36, 42, 44, 53, 54, 56–7t, 59t, 60, 60t, 70, 76, 78, 81, 108, 116, 121, 125t, 127, 134, 136, 137, 139*
  Department of Agriculture 54
  Gippsland 42, 84
  Melbourne 76, 120
  Shepparton 43
  Sunraysia 43
  Wodonga 43

Wallace, K.J. 25, 84, 86, 127, 151
Wasson, R.J. 78, 96, 97, 158
water
  eutrophication 39, **41–3**, 68, 69, 77
  ground water 44, 53, 70, 71t, 72, 75, 93, 100
  movements **89–93**
    aquifers 93, 94*
      perched 93
    canopy drip **91**, 99p
    canopy interception **89–91, 91t**
    capillary action, capillary fringe **100**
    drop impact 92, 94, 99p
    effects of vegetation change **93–5**
    ephemeral 93
    evaporation 93, 94, 100
    evapotranspiration 93
    groundwater recharge 97
    hydrology 100
    infiltration 39, 92, 97
    interception 89–91, 93, 94
    leaching 92
    overland flow 39, 92, 92t, **95**, 97, 99p, 100, 101, 132
    perennial 93
    rainwash 92, 95, 99p, 131
    seasonal 93
    stemflow **91–2**, 92t, 94, 97
    storage 94
    subsurface **97–100**
    surface wash (*see* rainwash)
    throughfall 91, 94
    throughflow **93**, 94*, 97
    transpiration 93, 97
  quality (*see also* pollution) 53, **69–75**
    chemical pollutants 70
    heavy metals 70, 71t
    loss of amenity 69, 70
    nutrient enrichment 39, **41–3, 44–5**, 69, 70, 71t
    pesticide residues (*see* pesticides)
    salinity (*see* salinisation)
    standards 44
    turbidity 69, 70, 71t
  resources, problems **71t**
  surface 42, 53, 70, 71t, 75
water catchment protection 76
  clearing bans 22, **76**
  Collie catchment 76
  Wellington dam 76
WAWRC (Western Australian Water Resources Council) 69, 158
weeds 4, 33–5t, **33–6**, 64, 65, 68, **85**
  Acacia nilotica 85
  annual ryegrass 36
  barley grass 36
  capeweed 36
  definition 33
  Mimosa pigra 33, 85
  noxious weeds 33
  partheneum weed 36
  pellitory 36
  prickly acacia 85
  rubber vine 33
  skeleton weed 85
  South African veld grass 36
  weed vectors 85
  wild oats 36
Wells, K.F. 20, 21, 158
West Australian, The 9, 47, 68

Western Australia 4, 13*, 14, 15t, 16, 18*, 20t, 21, 21t, 23t, 24, 25, 26, 27t, 32, 36, 44, 50*, 53, 54, 59t, 60, 60t, 62, 64, 66, 70, 73, 75, 76, 78, 80, 88, 92, 96, 108, 114, 116, 118, 125t, 137, 139*
Camballin irrigation scheme 67
Collie catchment 76
Jerramungup shire 69
Lake Argyle 42, 68
siltation 68
Merredin 114
Office of Catchment Management 136
Ord irrigation scheme 52, 67, 68
Peel Inlet 42
Pemberton 88t
Perth 35, 36, 44, 70, 86, 115t
Wellington dam 76
wheatbelt 24, 49, 91, 114p, 118, 119
Western Europe 9, 10, 62, 63*, 117, 141
France 63*
Germany 44, 63*, 106
Holland 129
Norway 129
Sweden 129
Western Farmer and Grazier 113
wetlands 32
wheat 31, 36, 66, 72t, 89, 91t
yields 62, 63*, 77, 79t
WHO (World Health Organisation) 44, 53, 54, 72t, 158
Wild, A. 89, 158
Williams, W.D. 118, 158
Williamson, D.R. 91, 94, 97, 155, 158
woodlands
brigalow 24
clearing **19–22**, 93
degradation **22–4**, 36
extent 19, 21, 21t, 23t, 24, 75
loss 20, 20t, 22
mallee 24, 91
wool industry 37
World Association of Soil and Water Conservation 124
World Bank 8, 9
World Commission on Environment and Development 139, 159
World Food Council 9, 159
World Health Organisation (*see* WHO)

Xanthorrhoea sp. 86